Engineering Works|

GW01393024

Contents

OXFORD

UNIVERSITY PRESS

1 What is engineering?

Before you start

1 Work with a partner. In your own language, think of four words to describe engineering. Compare your words with another pair and agree on four words. How do you say these words in English? Write down four English words to describe engineering.

Reading

2 Read the headings of the paragraphs (1–4) below. <u>Underline</u> any new words and look them up in the glossary or your dictionary. What do you think each paragraph will be about?

3 Read the paragraphs (1–4) below and check.

1 Engineering is everywhere

Almost everything we use in modern life is made by engineers. For example, if **a manufacturer** wants a faster car, a smaller personal stereo, or a better pen, they will ask a design engineer to find a practical **solution**.

2 Engineering is both theoretical and practical

Engineers use theory (ideas about engineering) to produce practical answers. The design solution must be a reasonable price, safe, and reliable. A new idea that is expensive, dangerous, or doesn't always work is not a good solution.

3 Engineers use a method

Generally, engineers solve problems in a **methodical** way. Engineers:
1 **define** the problem,
2 **design** a solution,
3 test the solution,
4 **evaluate** the solution.
If the solution isn't right, the process is repeated. When a good solution is found, the next step is to:
5 communicate the solution.

4 Anyone can use engineering ideas

This method of problem-solving is useful in **everyday** life. For example, you can use the five steps next time you prepare for a test.
1 Define the problem: I want to pass my test next week.
2 Design a solution: I will study for three hours a day.
3 Test the solution: Study for three hours a day and take the test.
4 Evaluate the solution: Have I passed the test with a good mark? Yes = a good solution. No = a bad solution, so think of a better one.
5 Communicate the solution: Tell your friends about your test-passing technique.

4 Read the four paragraphs again and decide if the sentences (1–4) below, are true (T) or false (F).

1 Lots of things are made by engineers.   T/F
2 Engineering isn't practical.   T/F
3 Engineers must think carefully.   T/F
4 Only engineers can solve problems.   T/F

Vocabulary

5 Match the **highlighted** words from the text with the meanings (1–7) below.

1 plan   5 careful
2 say exactly   6 assess the success of
3 a business   7 normal
4 answer

Writing and Speaking

6 Read the paragraph headings again. Do you agree with them?

7 Work with a partner. Choose one of the problems below or your own problem. Solve it using the five steps. Make notes.

- You want to go away for a weekend with your friends but your parents want you to study.
- You want to buy a CD player but you haven't got any money.

8 Explain your problem and the solution to another pair of students.

9 Look at the four words you wrote to describe engineering at the beginning of the unit. Do you want to change them?

▶ Get real

Find an example of a new, improved design, for example, a new model of car, household appliance, or personal stereo. Compare the original and the new one. Which features are different? Is the new one better?

2 The right person in the right job

Before you start

1 What sort of person are you? First, answer the questions below. Then compare your answers with the rest of the class. Do you all like the same things?

In your spare time, do you prefer:

- being alone or with other people?
- being inside or outside?
- being busy or relaxing?
- playing sport or watching television?
- wearing smart or casual clothes?

Reading

2 Read the text *A job in Engineering*. Put the main ideas (A–D) in the same order as they are in the text.

A You need to think carefully about your personality.
B There are lots of different jobs in engineering.
C Think carefully about what you are interested in.
D Engineering is a big subject.

A job in Engineering

1 There are lots of different types of engineering. The one thing they have in common is that they all use Maths and Science to improve industry and manufacturing. The whole science of engineering can be broadly divided into three main areas:
- civil engineering (buildings, roads, etc.)
- mechanical engineering (machines, including tool-making)
- electrical engineering (electricity, lighting, etc.)

2 Each of these three main area can be divided again into specialist subjects: civil engineering covers mining and bridge building, mechanical engineering covers aeronautical and automobile engineering, electrical engineering covers electricity generation and wiring.

3 Clearly there is a big difference between building a road and designing a computer system so the best advice for students is:
- think carefully about which area of engineering interests you most. It is difficult to study if you are not interested – and you may do the job until you are 60 years old.
- think about what sort of person you are. Will you be happiest working in an office, in a factory, or outdoors? Do you mind getting dirty? Do you want to work with other people or alone? If you like wearing high heels and beautiful clothes, you may not be happy on a building site.

4 When you have decided which area you are interested in and thought realistically about what sort of person you are, then you can decide what sort of engineer you want to be.

3 Read the text again. Choose the correct words to complete the sentences (1–4) below.

1 Engineering *is/isn't* a small area.
2 Engineering *is/isn't* about Science and Maths.
3 Office buildings and bridges *are/aren't* examples of civil engineering.
4 Tool- and machine-making *are/aren't* examples of electrical engineering.

Vocabulary

4 Look at the types of engineers (1–5) below. First, <u>underline</u> any new words and check the meaning in the glossary or your dictionary. Then write whether the jobs are *indoor* or *outdoor*, and *dirty* or *clean*.

1 petroleum
_____

2 sanitation
_____

3 textile
_____

4 computer
_____

5 chemical
_____

Writing and Speaking

5 Write your name and the type of engineer you want to be on a small piece of paper. Put your paper in a box and take out another student's paper. Write three questions to check if that person is choosing the right type of engineering.

For example:
Mining engineer
- *Do you like working indoors or outdoors?*
- *Do you mind getting dirty?*
- *Do you like going underground?*

6 Find the student and ask them the questions. Have they chosen the right type of engineering?

> ▶ **Get real**
> Find examples of job advertisements for engineers in your town. What type of engineers are employers looking for? Find out the English words.

Before you start

1 First, answer these questions about your studies. Then discuss your ideas with the rest of the class.

- Which subjects do you study? Are there any subjects you would like to drop?
- Is your course practical? Do you like this way of working?
- How are you assessed? Do you think this is fair?

Reading

2 Read the text quickly and choose the correct answers to questions 1–3 below.

1 Where is the text from?
   a a textbook **b** a leaflet

2 Who is the information for?
   a new students **b** teachers and parents

3 What is the text about?
   a one course **b** lots of courses

Who is the First Diploma for?

It is a foundation course for students with a general interest in engineering. You will learn about the different types of engineering; you do not specialize in one area.

What qualifications do I need?

You must be at least sixteen years old with an interest in Engineering. You need at least three GCSEs including Mathematics, Science, and Design and Technology.

What will I learn on the course?

You will learn:
- practical skills in manufacturing and maintenance
- about engineering materials, Computer Assisted Design (CAD), engineering measurement
- key skills in Information Technology

How will I learn?

The course is full-time for one year.
You will spend some time in the classroom but most of your time will be spent doing practical tasks in the workshop, in the laboratory, or on computer screen.

How will I be assessed?

Each project is marked (continuous assessment) and there are tests at the end of each term.
What can I do when I finish the course?
You can use your First Diploma to help you to find a job as an apprentice. Alternatively, you can continue your studies and specialize in the area that interests you most.

---

3 Read the text again and decide if the sentences (1–6) below are true (T) or false (F).

1 The First Diploma is a beginner level course. T/F
2 Students learn general things about engineering. T/F
3 The course isn't practical. T/F
4 Students take one big exam at the end of the year. T/F
5 At the end of course, you can apply for a job as a trainee. T/F

Vocabulary

4 Complete the definitions (1–6) below with the highlighted words in the text.

1 To _____ means to know a lot about one part of a subject.
2 _____ means all your work on the course is part of the final mark.
3 _____ are the most important things to learn.
4 A _____ teaches you general things about a subject.
5 An _____ is someone who works with an experienced person to learn the job.
6 You get _____ when you pass exams.

Writing

5 Answer the questions (1–6) about *your* course and write a leaflet for it.

1 Who is the course for?
2 What qualifications do I need?
3 What will I learn on the course?
4 How will I learn?
5 How will I be assessed?
6 What can I do when I finish the course?

▶ ### Get real

Find a leaflet for your school or college, or a course at a college you are interested in. Compare it with the one you wrote. Can you improve either of them?

4 The course for you

Before you start

1 Ask other students in your class the questions below. If they answer *yes*, ask more questions.

Have you ever …
… had an interview?
… filled in an application form?
… written to someone asking for information?

For example
Have you ever had an interview?
Yes, I have.
What for? What happened?

Reading

2 Read the text quickly and choose the correct answers to questions 1–3 below.

1 What is the leaflet about?
  a Engineering courses at Coalport Technical College
  b All courses at Coalport Technical College
2 What does the leaflet describe?
  a lots of courses   b one course
3 How can you get more information?
  a by telephoning   b by returning the form

3 Read the text again and match the people (A–E) below with a suitable course. One person isn't suitable for any of the courses.

Person A finished Level 1 last year. He wants to work as a welder on oil rigs.
Person B got her exam results last week. She passed Maths, Design and Technology, English, and History. She wants to work in design.
Person C works in her father's garage. She hasn't passed any exams but she is good at mending cars and wants to return to studying.
Person D left school in 2000 with no exam passes. Since then he has worked in a jeans shop and a hamburger café.
Person E has five GCSEs and wants to work as a telephone engineer.

Vocabulary

4 Complete the definitions (1–6) below with the highlighted words in the text.

1 An _____ is a person who wants a job or a place on a course.
2 An _____ is a formal meeting.
3 To _____ is to make something ready.
4 _____ are ways of sending information, news, etc. from one place to another.
5 _____ is joining metal by heating.
6 _____ means putting machinery in place.

ENGINEERING COURSES

at Coalport Technical College, Blackstock

All courses are taught at Coalport Technical College, Blackstock and can be studied full-time or part-time. The minimum qualification for a place on a Level 2 course is four GCSEs or a Level 1 Certificate.

LEVEL 1 Certificate in Engineering

This course teaches basic, key skills. It is suitable for students who left school early or have no qualifications. Selection will be based on the applicant's work experience and an interview.

LEVEL 2 Certificate in Electrical and Electronic Engineering

This course prepares students for jobs in radio and electronic communications.

LEVEL 2 Certificate in Fabrication

This course prepares students for jobs in welding, sheetmetal work, and general engineering.

LEVEL 2 Certificate in Mechanical Engineering

This course prepares students for a wide range of jobs including machining, fitting, tool-making, CAD and CAM.

✂

For more information please visit our website or return this form to the departmental secretary.

Name _____
Address _____
Telephone/e-mail _____
Date of birth _____

Please send me details of:
LEVEL 1 Certificate in Engineering ☐
LEVEL 2 Certificate in Electrical and Electronic Engineering ☐
LEVEL 2 Certificate in Fabrication ☐
LEVEL 2 Certificate in Mechanical Engineering ☐
I am interested in full-time study ☐ part-time study ☐

Speaking and Writing

5 Ask your partner the questions on the form at the end of the leaflet.

6 Complete the form for your partner.

▶ **Get real**
Use the Internet to find college information or contact a college in your area and ask for leaflets about their courses. Is the information helpful? How do you get more details or apply for a course?

5 | What is it made from?

Before you start

1 Look at the objects (a–d) below. Explain which is the best material *glass*, *plastic*, or *metal* for each one.

a a fork
b a football
c a window
d a bicycle

Reading

2 Put each heading into the correct column (A, B, or C), in the table below.

Uses ■ Properties ■ Material

3 Read the information in the table and find out which material (1–10) is best for:

a water pipes
b a knife for cutting a microscope lens
c connecting a socket to the electricity supply
d a bicycle frame
e television casing

	A	B	C
1	aluminium	light, easy to shape	aircraft, window and door frames, cooking foil
2	brass (copper and zinc)	doesn't rust in contact with air and water, strong	valves, taps
3	cement	mixed with water it dries to a hard material	pre-made building blocks, to hold bricks together
4	copper	easily made into wire, carries electricity well	electrical wire, tubing
5	diamond	hardest natural material, can cut glass and metal	industrial cutting and grinding
6	glass	clear, hard, breaks easily	windows, bottles
7	iron	hard	engineering
8	mild steel (iron + 0.15–0.3% carbon)	hard, strong, quite easy to shape	bridges, ships, cars
9	optical fibre	carries light and coded messages	lighting, cable TV, telecommunications
10	plastic	light, strong, easy to shape	hard hats, telephones, boats, computer casing

Vocabulary

4 Match the properties from the table (1–6) with their opposites (a–f). Use the glossary or your dictionary to help you.

1 breaks easily	a heavy
2 clear	b tough
3 easy to shape	c opaque
4 hard	d rigid
5 light	e weak
6 strong	f soft

5 How do you say these materials in your language?

nylon ■ cardboard ■ asphalt

Writing

6 Choose a material from Exercise 5 and write a description of its properties and uses.

7 Exchange descriptions with your partner and identify their material.

Speaking

8 Look at the list of materials in Exercises 3 and 5 again. Discuss the questions (1–3) below. Check any words you don't know in the glossary or your dictionary.

1 Are any of the materials found or manufactured in your country?
2 Which materials can be recycled?
3 Does the manufacture or disposal cause environmental problems?

▶ ***Get real***
What material did Charles Goodyear develop? What is it used for?

6 Bend it like ...

Before you start

1 Read the sentences (1–3) below. What does the word *smart* mean in each one (*clever, fashionable*, or *formal*)?

1 He wore a *smart* suit to the meeting.
2 She's the *smartest* girl in her class.
3 They stayed in a *smart* hotel in New York.

Reading

2 Look at the title of the text, *Smart materials*. Do you think the materials are clever, fashionable, or formal? Read the text and check.

SMART Materials

Smart – or shape memory – materials are an invention that has changed the world of engineering. There are two types: metal alloys and plastic polymers. The metal alloys were made first and they are usually an expensive mixture of titanium and nickel.

Shape memory materials are called 'smart' because they react to changes in their environment, for example:

- plastics that return to their original shape when the temperature changes. One use is in surgery where plastic threads 'remember' the shape of a knot, react to the patient's body temperature and make themselves into stitches.
- metal alloys that have a 'memory' and can return to their original shape. They are used in medical implants that are compressed so they can be put inside the patient's body through a small cut. The implant then expands back to its original shape. More everyday uses are for flexible spectacle frames and teeth braces.
- solids that darken in sunlight, like the lenses in some sunglasses.
- liquid crystals that change shape and colour. These have been used in climbing ropes that change colour if there is too much strain and weight on them.

The future of these materials and their possible uses is limited only by human imagination. One clever idea is that if cars were made of smart metal, a minor accident could be repaired by leaving the car in the sun!

3 Read the text again and choose the correct answers for questions 1–4 below.

1 Smart materials change when
   a the weather changes.
   b something affects them.
   c the light is switched on.

2 Plastic threads are used for
   a sewing.
   b stitching.
   c knitting.

3 Medical implants made from shape memory alloys are good because
   a they save lives.
   b they change colour.
   c they are easy to put in.

4 Climbing ropes with liquid crystals change colour to
   a warn you.
   b amuse you.
   c make you heavy.

Vocabulary

4 Complete the definitions (1–8) below with the highlighted words in the text.

1 An _____ is something medical put inside the body, e.g. a heart valve.
2 You need a good _____ to think of new and interesting ideas.
3 The _____ is the first or earliest.
4 _____ are materials made from mixing two metals.
5 To _____ means to become bigger.
6 To _____ is to change because something else happens.
7 The _____ is everything around a person or thing.
8 To be _____ means to be made smaller.

Speaking

5 Work with a partner. Choose one of the smart materials in the text. Think of five interesting ways it could be used. Compare your ideas with other students. How many original ideas are there in your class?

> ### Get real
> Think of examples from nature that are like smart materials, for example, things that can change shape or colour, or repair themselves.

7 A picture is worth a thousand words

Before you start

1 What does the title of this unit mean? Is there a similar expression in your language?

2 Think of situations in engineering where a picture is more useful than words. Discuss your ideas with the rest of the class.

Reading

3 Read the sentences below. What is the difference between 2D and 3D?

- An image on paper, for example a diagram or photograph, is two dimensional (2D).
- A model or statue is three dimensional (3D).

4 First, look at the pictures. What do you think this text will be about? Then read the text and check.

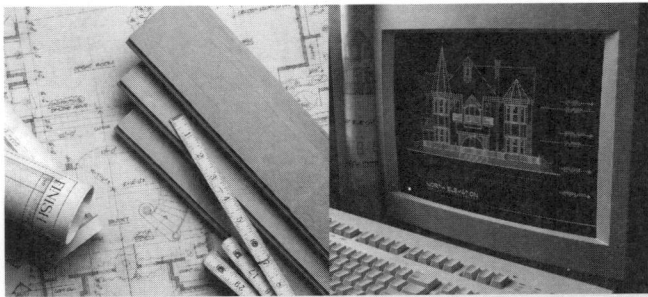

A In the past, technical drawings for industry and architecture were drawn by hand, i.e. people worked at drawing boards with drawing equipment. These hand-drawn diagrams provided clear technical information but were slow and expensive to make. Nowadays, working drawings are done on computers, which is much quicker.

B Computers can also:
1 save, change, and recycle the drawings
2 make 3D images
3 make drawings bigger or smaller
4 keep an electronic library of standard parts
5 make symmetrical images of components
6 make accurate and consistent drawings

C A good way to explain the advantages is to think about architectural drawing. Features such as windows and doors can be moved until the architect likes the building. Images of the rooms are created in 3D so the viewer can 'walk' through the rooms. Designers can also experiment with different arrangements of furniture and colours.

5 Read the text and choose the correct answers to questions 1–4 below.

1 What is the text about?
   a Computer assisted design
   b Working with computers

2 What is paragraph A about?
   a The history and future of CAD systems
   b The connection between technical drawing and CAD

3 What is paragraph B a list of?
   a The problems of using computers in design
   b The advantages of using computers in design

4 What does Paragraph C describe?
   a How CAD is used in designing machines
   b How CAD is used in designing homes

6 Read paragraph B again. Match each point (1–6) in the text with a benefit from the list (a–d) below.

   a You can draw 50%, then make a mirror image.
   b You don't waste time drawing things again and again.
   c You make fewer mistakes.
   d You see the finished shape in 3D.

7 Explain, in your own language, why the word 'walk' is in inverted commas in paragraph C.

Speaking

8 Work with a partner. Decide which is the biggest benefit of using a computer for technical drawing. Can you add other benefits to the list? Compare your ideas with the rest of the class.

Vocabulary

9 Complete the definitions (1–8) below with the highlighted words in the text.

1 _____ means having two halves the same shape and size.
2 _____ means made by a person.
3 _____ means to use something again.
4 A _____ is a person looking at something.
5 _____ means correct in every detail, with no mistakes.
6 _____ are pictures or drawings.
7 An _____ is a person who designs buildings.
8 _____ means always the same.

▶ ***Get real***
Find out about other images made by computers. For example, how are computer images used on television and in films?

8 Getting away from it all

Before you start

1 Have you visited a company or factory as part of a field trip? Where did you go? Did you find out anything interesting? Where would you like to go? Why?

Reading

2 Read the text and decide if the sentences (1–6) below are true (T) or false (F).

The speaker…

1 …is a teacher.	T/F
2 … is talking about a factory system.	T/F
3 …is talking about jobs in the factory.	T/F
4 …is talking to other engineers.	T/F
5 …is talking to students.	T/F
6 …is a worker in the factory.	T/F

3 Read the text again and match the paragraphs (1–5) with the topics (A–E) below.

A Background information
B Rules for the factory visit
C Benefits of the system
D Welcome
E Basic information about their system

4 Read the text again and write short answers to these questions.

1 Are CAM and CNC the same?
2 Is manufacturing the second stage in their process?
3 Can CNC operate other machines?
4 Does the speaker like the system?
5 When will he/she answer the students' questions?

5 In your own language, write down the three advantages of the system mentioned in the text. Can you add any more ideas?

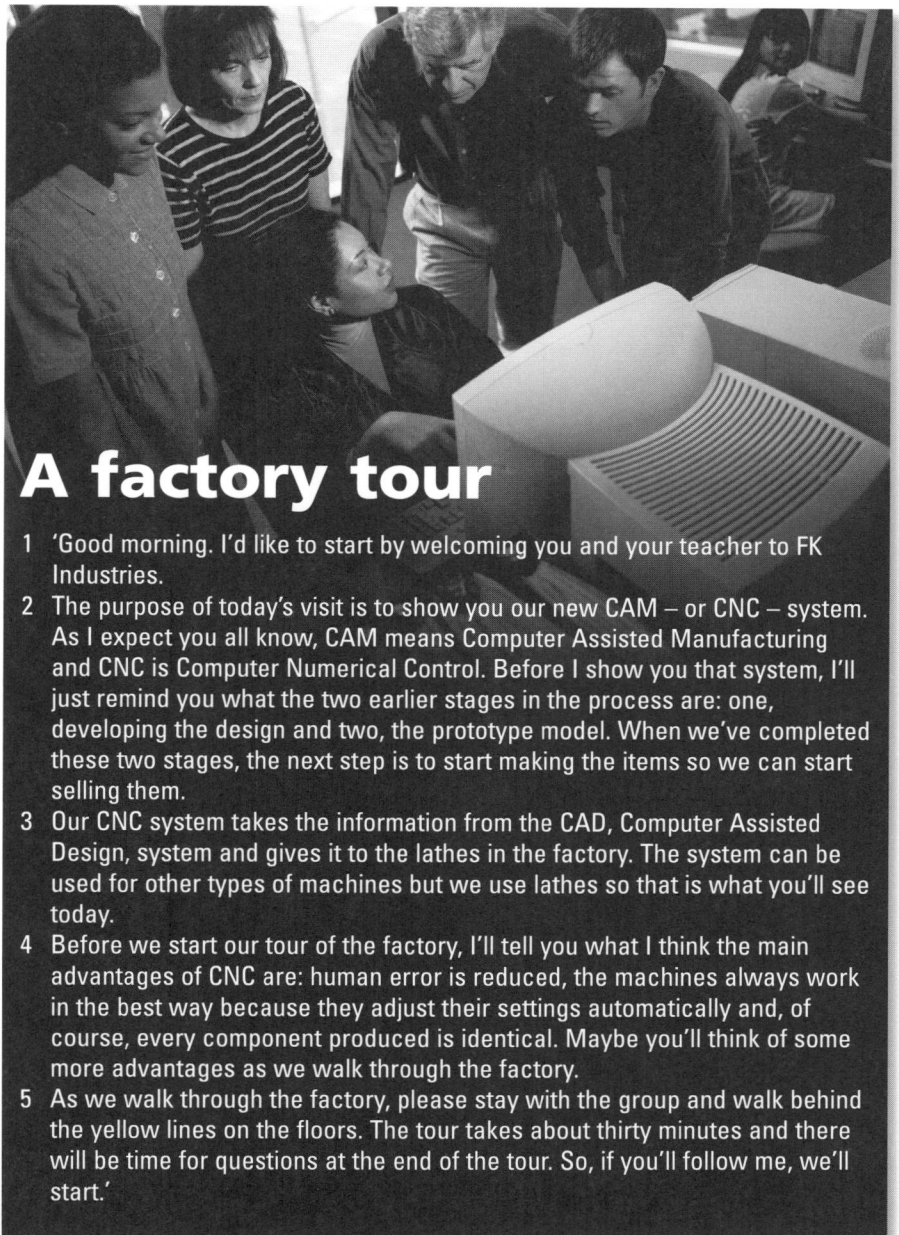

A factory tour

1 'Good morning. I'd like to start by welcoming you and your teacher to FK Industries.
2 The purpose of today's visit is to show you our new CAM – or CNC – system. As I expect you all know, CAM means Computer Assisted Manufacturing and CNC is Computer Numerical Control. Before I show you that system, I'll just remind you what the two earlier stages in the process are: one, developing the design and two, the prototype model. When we've completed these two stages, the next step is to start making the items so we can start selling them.
3 Our CNC system takes the information from the CAD, Computer Assisted Design, system and gives it to the lathes in the factory. The system can be used for other types of machines but we use lathes so that is what you'll see today.
4 Before we start our tour of the factory, I'll tell you what I think the main advantages of CNC are: human error is reduced, the machines always work in the best way because they adjust their settings automatically and, of course, every component produced is identical. Maybe you'll think of some more advantages as we walk through the factory.
5 As we walk through the factory, please stay with the group and walk behind the yellow lines on the floors. The tour takes about thirty minutes and there will be time for questions at the end of the tour. So, if you'll follow me, we'll start.'

Speaking and Writing

6 With a partner, think of three questions to ask at the end of the factory tour. Compare your ideas with another pair. Can any of you answer the questions?

7 You are going to take a group of English visitors around your school. Prepare what you will say to them. Write notes, practise what you will say, and ask a friend to listen to you. Revise your notes if necessary.

> ### ▶ Get real
> 1 Look at Exercise 4 again. Are there any unanswered questions? Can you find out the answers?
> 2 Which international companies do work you are interested in? Find their website or their postal address and e-mail or write to them for information about their products.

Before you start

1 In your own language, write down three reasons *for* and three reasons *against* owning cars. Find out the English words and discuss your ideas with the rest of the class. Which students do you agree with?

Reading

2 Read 'The Mini Story'. Which three paragraphs are from the same newspaper article and which one is from a fashion magazine?

3 Put the three paragraphs from the newspaper article in the correct order.

Vocabulary

4 Find words in the text that mean:

1 employees who work machines (paragraph A)
2 factory (paragraph C)
3 very modern (a phrase, paragraph C)
4 working conditions (paragraph A)
5 where the cars are put together (two words paragraph A)

5 Read the text again and answer the questions (1–5) below.

1 When was the first Mini made?
2 Why is the Mini factory in Oxford special?
3 How many new Minis are made each year?
4 How many people work at the Mini factory?
5 Why is the new factory better for the workers?

THE MINI STORY

A There are 2,500 employees at the plant and the working environment is good. The car assembly line is designed ergonomically to be easy to use and comfortable for the operators. For example, the car is raised, lowered, and turned through 90 degrees so the workers can do their jobs comfortably and easily. Old-fashioned, noisy, compressed-air tools have been replaced with quieter and more accurate electric tools.

B The first Mini was first made in 1959 and since then over five million people have owned one. BMW, a German car manufacturer, now owns the Mini and the newest model is being manufactured at an advanced production system in Oxford, England.

C During the 1990s approximately £500 million was spent to change an old Oxford car factory into a state-of-the-art manufacturing plant. The Oxford plant now produces around 100,000 Minis a year.

D In Britain in the 1960s the only really cool car was the Mini. Everybody wanted one. It starred in advertisements and films and was as famous as the Beatles or the Rolling Stones. Anyone who was young, rich, famous, and fashionable had to be photographed sitting on, in, or just near one. And anyone who was poor, unknown, and not very fashionable wanted one too. They were small and cheap and suited the mood of the post-war generation who had more money and freedom than their parents had ever had.

Writing and Speaking

6 In your own language, add to this list of the things people consider when they choose a car. Then find the English words.

comfort ■ image ■ fuel consumption ■ …

7 Work with a partner and choose a suitable car for these people. Tell the class what sort of car you have chosen and why.

- a family
- a young, single person
- a film star
- a business person

For example:
A Seat Ibiza is a good car for a young, single person because it's small, cheap to buy, and has good fuel consumption.

▶ *Get real*
   Find out what connects the Mini and the 1969 film, *The Italian Job.*

10 Made by hand

Before you start

1 What are the advantages and disadvantages of each option below? Discuss your ideas with the rest of the class.

- studying in a small/big school
- living in a village/city
- working in a family business/an international company
- shopping in a local shop/a large supermarket

Reading

2 First, look at the words in the box. Check the meaning of any new words in the glossary or your dictionary. Then complete the text 'A handmade car' by putting one word in each space. Use the words in the box.

> craftsmen ■ highly-skilled ■ skills
> ■ traditional ■ unique

3 Read the text again and decide if the sentences (1–5) below are true (T) or false (F).

1 The Morgan is made by machines.   T/F
2 Old and new ideas are used to make Morgans.   T/F
3 Morgan cars aren't made on an assembly line.   T/F
4 Morgan engines are old-fashioned.   T/F
5 You can walk into the Morgan factory, buy a car and drive it home.   T/F

Vocabulary

4 Read the texts about the Morgan and the Mini (in Unit 9) again. Put the words in the box into the table below. Some words may fit in more than one column.

> boring ■ classic ■ difficult ■ easy ■ fashionable ■
> interesting ■ modern ■ organized ■ peaceful ■
> requires expertise ■ requires patience ■
> requires skill ■ requires you to work quickly ■
> traditional

The Mini	Jobs in the Mini factory

The Morgan	Jobs in the Morgan factory

A handmade car

The Morgan is a (1) _____ car: it is made in Britain by a family-owned company and it is handmade.

Each Morgan is made individually. Modern materials and up-to-date manufacturing technology are combined with 100-year-old (2) _____. There are no assembly lines because each stage of the manufacturing is done by (3) _____ craftsmen. For example, the wooden frame is made in the same way as the first Morgan in 1909, upholsterers make the leather seats, and sheet metalworkers make the panels by hand.

In contrast to all these (4) _____ skills, Morgan engineers make precision mechanical components using modern Computer Numerical Control (CNC) machinery so a Morgan driver has a state-of-the-art engine in a traditionally-made car.

It takes a *long* time to make a car by hand. The Morgan factory produces about 500 cars a year. Buyers put their name on a waiting list and then wait for the factory to tell them that their car is finished. The shortest wait is about two years – and sometimes the wait is five years. Like proud parents-to-be, people on the waiting list can visit the factory to see their car being made and to talk to the (5) _____ doing the work.

Speaking

5 Explain why it takes longer to make a Morgan than a Mini.

6 Discuss these questions:

- Which car would you rather own and why?
- Which factory would you rather work in and why?

▶ **Get real**

Find out which are the cheapest and most expensive cars in your country. Are they made locally or imported? How are they made?

Before you start

1 Complete the statement below for you. Then compare your answer with the rest of the class. Which is the most popular way to learn?

When I learn to do something practical, I prefer …

a … to see someone demonstrating it.
b … someone to help me do it.
c … to follow a diagram.
d … to try and ask for help if things go wrong.

Vocabulary

2 Label the computer components on the diagram below. Use the words in the box.

> cable ■ computer ■ keyboard ■ monitor ■ mouse ■ printer

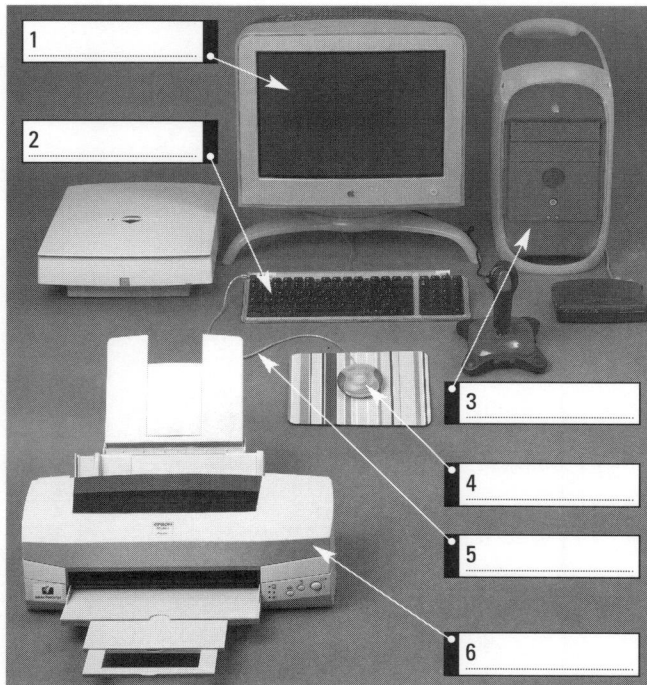

3 Complete the sentences (1–5) below. Use the verbs in the box. There is one extra verb that you do no need to use.

> connect ■ disconnect ■ loosen ■ plug in ■ tighten ■ unplug

1 If you don't pay the bill, the electricity company will _____ the supply.
2 _____ the screws before you take the plug out.
3 It's sensible to _____ your computer if there is a bad storm.
4 If you don't _____ the TV, it won't work!
5 _____ the video cable to the TV.

Reading

4 Read the instructions (1–6) below and match them with the diagram.

1 Connect the keyboard cable to the back of the computer. *c*
2 Connect the mouse cable to the back of the computer.
3 Plug in the monitor cable; be careful not to bend the pins.
4 Tighten the screws.
5 Connect the speakers to the back of the computer.
6 Plug the computer, monitor, and speaker cables (in that order) into the mains supply.

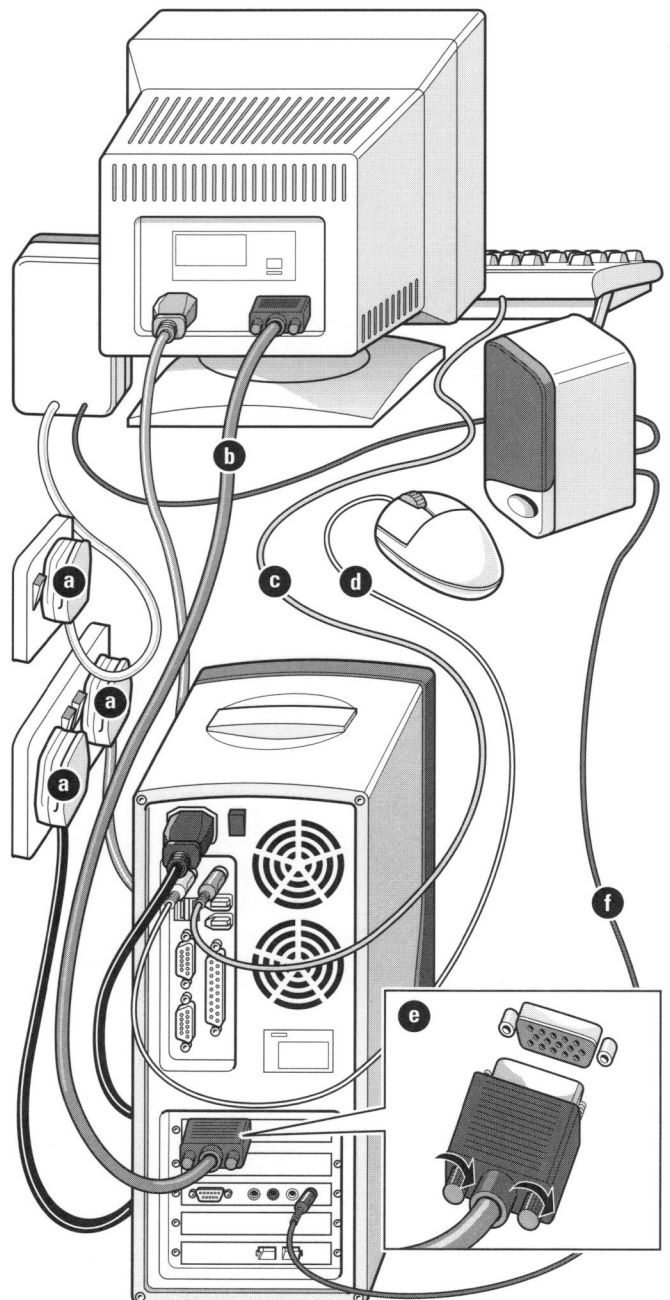

5 Read the instructions below and match the spoken instructions (1–3) with the written instructions (a–c).

1 Put in some water and turn on the gas.

2 If I were you, I'd put the bulb in first.

3 You press that and the back opens.

a Refer to diagram 1. Button A releases the locking mechanism.

b Remove the lid and fill from the cold tap. Place the kettle on the centre of the gas ring making sure that it is stable before turning on the gas.

c Insert a 60W bulb *before* putting the plug in the socket.

6 Read the instructions in Exercise 5 again. Which instructions are about:
- a new kettle?
- a new desk lamp?
- a new camera?

7 What are the differences between written and spoken instructions? Can you explain the differences?

Writing

8 Read these instructions for connecting a DVD to a TV. Are they clear? Explain why/why not.

I think you join the TV to the DVD with that thing. Get the other thing and put one end in there. The other end goes in the DVD.

9 Complete the instructions for connecting a DVD recorder to a TV set. Use the words in the box.

aerial ■ cable ■ mains ■ plug ■ socket ■ TV

INSTRUCTIONS FOR CONNECTING THE DVD TO THE TV

Switch off your TV set.

Remove the aerial cable (**1**) _____ from your TV set. Insert it into the ANTENNA socket at the back of the DVD recorder.

Insert one end of the aerial (**2**) _____ into the TV socket at the back of the DVD recorder and the other end into the (**3**) _____ input socket at the back of the (**4**) _____ set.

Plug a special scart cable into the scart (**5**) _____ at the back of the DVD recorder and the corresponding scart socket at the back of the TV set.
Switch on the TV set.

Insert one end of the supplied mains cable into the (**6**) _____ socket at the back of the DVD recorder and the other end into the wall socket.

▶ **Get real**
Find some instructions for an electrical appliance, for example a stereo system. How many of the components do you know in English?

Before you start

1 Where would you find the notices below? How do you say these things in your language?

  a Don't lean out of the windows.
  b Now wash your hands.
  c Beware of the dog.

2 Look at the picture. What safety equipment is the operator wearing?

3 Look at the signs inside the back cover. Match the meanings (1–4) with the shapes (a–d) and colours (e–h).

meaning
  1 you mustn't do this
  2 you must do this
  3 there is a danger
  4 this material is dangerous

shape
  a triangle
  b circle with diagonal line
  c square
  d circle

colour
  e yellow and black
  f red and white
  g blue and white
  h orange and black

4 What do the signs mean? Where would you find them? Discuss your ideas with your class.

Reading

5 Read the instructions and warnings (1–11) below. First, underline any new words and check the meaning in the glossary or your dictionary. Then match the sentences with the signs (a–k) inside the back cover.

  1 Be careful.
  2 Beware of industrial vehicles.
  3 Don't smoke here.
  4 Don't walk here.
  5 Risk of death.
  6 This material is corrosive.
  7 This material is explosive.
  8 This material is flammable.
  9 Wear a hard hat.
  10 Wear ear defenders.
  11 Wear goggles to protect your eyes.

Vocabulary

6 Choose the best word to complete the sentences (1–5) below.

  1 Petrol and oil are *flammable/vehicles*.
  2 Acid is *flammable/corrosive*.
  3 TNT and dynamite are *corrosive/explosive*.
  4 Wear *a hard hat/goggles* when you work with chemicals.
  5 You must wear *a hard hat/goggles* on a building site.

Writing and Speaking

7 Work in pairs. Choose one of the places in the box, or another place. Write two rules, one thing you *must* do and one thing you *mustn't* do.

> computer room ■ engineering workshop ■ school sports hall ■ your bedroom

For example,
In a school sports hall you mustn't wear outdoor shoes and you must return equipment.

8 Design and draw a sign for each rule.

9 Show your signs to another pair. Can they work out the meaning?

▶ ## Get real

Use the Internet, reference books, or leaflets to find out the answer to these questions.
  a) What are the standard European colours for health and safety signs?
  b) Find five examples of warning signs in your own language. Translate them into English. Can your partner translate them back into your language?

13 Are you sitting comfortably?

Before you start

1 What is repetitive strain injury (RSI)? What is the equivalent phrase in your language?

Reading

2 Read the text and match the headings (A–D) with the paragraphs (1–3). There is one extra heading that you do not need to use.

A Advice for computer workers   C General advice
B Advice for factory workers   D What is RSI?

Repetitive Strain Injury (RSI)

1

Any person who repeats the same movement a lot of times can develop repetitive strain injury. Factory workers, computer operators, sports people, and musicians are at the most risk because their jobs involve making the same movement thousands of times. The symptoms of RSI include: pain and/or burning in the damaged area, difficulty in moving, and loss of feeling.

2

It is difficult to cure RSI but you can avoid it before it starts. To prevent RSI, workers at risk should:
- take regular breaks from their work to stretch and move about
- learn to sit and move correctly so they use their bodies naturally

3

People who use computers for a long time have a high risk of developing RSI. Here are some basic rules for working safely at a computer:
- take regular breaks to stretch and relax
- move the screen to eye level or a little bit lower
- don't hold the mouse for too long or too tightly
- sit with your back relaxed, shoulders down and your neck straight
- keep your wrists relaxed, your elbows at about 90 degrees and the lower parts of your arms parallel to the desk top
- use an adjustable chair
- keep your feet flat on the floor

Vocabulary

3 Match the highlighted words in the text with the meanings (1–6) below.

1 a danger
2 can be moved into different shapes or positions
3 signs of an illness
4 stop something happening
5 to make an illness better
6 without tension or strain

4 Look at the diagram and put the labels (a–g) below in the correct place.

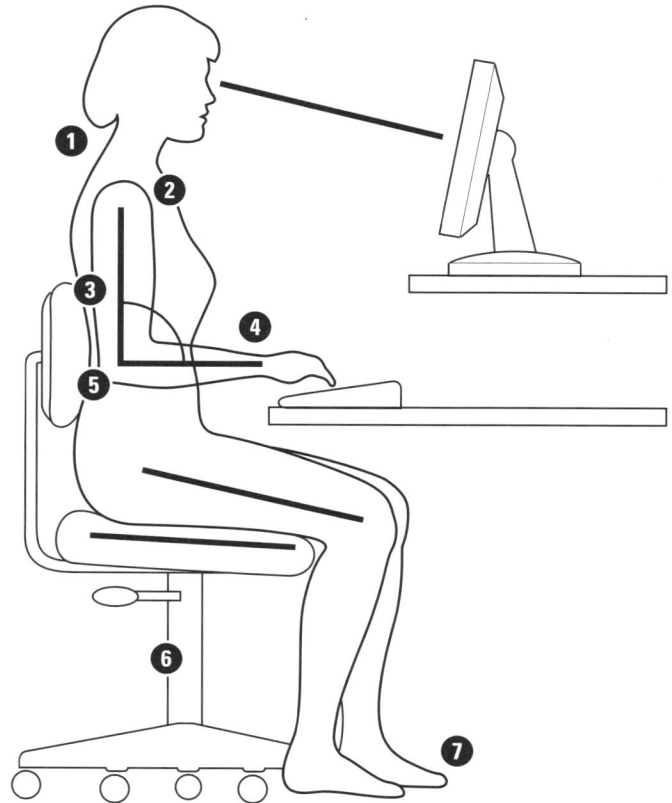

a elbows at 90 degrees
b feet flat on the floor
c head and neck straight and relaxed
d lower arm horizontal
e shoulders down
f upper arm vertical
g use an adjustable chair

Speaking and Writing

5 Work in pairs. Discuss the risks and sensible safety precautions for another activity. You can use your own ideas or choose one of these. Then compare your ideas with the rest of the class.

- using a mobile phone
- cycling
- using a welding torch

6 Make a safety information leaflet about the topic you chose in Exercise 6. Write a list of five things the user should do. Draw and label a diagram to make the information clear.

> ▶ **Get real**
> Find a safety information leaflet or advice on the Internet about a different health and safety issue. How good do you think the advice is?

14 Small is beautiful

Before you start

1 How are these things carried from one place to another?

   a electricity   **b** radio signals   **c** gas

Reading

2 Read the text quickly and choose the best title, A, B, or C.

   A The history of cabling and telecommunications

   B A short introduction to optical fibres

   C Uses of glass in industry and technology

@ Optical fibres

Address: **@** Optical fibres

Optical fibres started to replace some uses of copper cables in the 1970s. They are made from glass and are usually about 120 **micrometre**s in **diameter**. Some of the most common everyday uses are in telecommunications, close-circuit television (CCTV), and cable television.

1 _____
Optical fibres carry signals more **efficiently** than copper cable and with a much higher bandwidth. This means that fibres can carry more channels of information over longer distances.

2 _____
Optical fibre cables are much lighter and thinner than copper cables with the same bandwidth. This means less space is needed in underground cabling **ducts**.

3 _____
It is difficult to steal information from optical fibres. They are not harmed by electromagnetic interference, for example from radio signals or lightning. They don't **ignite** so they can be used safely in **flammable** atmospheres, for example in petrochemical plants.

4 _____
Optical fibres are more expensive **per** metre than copper. However, one optical fibre can carry many more signals than a single copper cable and the longer transmission distances mean that fewer expensive repeaters are required. Also, copper cable uses more electrical power to deliver the signals.

5 _____
Optical fibres can't be spliced as easily as copper cable. Employees need special training **to handle** the expensive **splicing** and measurement equipment.

Internet zone

---

3 Read the text again and match the headings (A–E) with the paragraphs (1–5).

   A Training and skills   **D** Price

   B Size and weight   **E** Capacity

   C Security

4 Which paragraphs describe *advantages* of optical fibres and which describe *disadvantages*?

Vocabulary

5 Complete the definitions (1–9) below with the **highlighted** words in the text.

   1 A _____ is one millionth of a metre.

   2 The _____ is the distance across a circle.

   3 A _____ substance is one that burns easily.

   4 _____ means joining the ends of two cables together.

   5 To _____ means to start to burn.

   6 _____ are tubes for carrying cables.

   7 _____ is a common short way of saying 'for each'.

   8 _____ means to touch with your hands.

   9 _____ means in a way that produces a good result and doesn't waste time, energy, or resources.

Writing and Speaking

6 Write two advantages and two disadvantages of using optical fibres instead of copper cable. Compare your ideas with the rest of the class.

▶ ***Get real***
Use the Internet, magazines, or newspapers to find out about another interesting invention of the last twenty years. Try to find out two advantages and two disadvantages of the invention and tell your class. Decide who found out about the most interesting invention.

15 Big is best

Before you start

1 Work in pairs. You have one minute. How many different dams or tunnels can you think of? Compare your answers with the rest of the class.

Reading

2 Read the text quickly and decide which structure it describes.

  a The Hoover Dam
  b The Arlberg Tunnel
  c The Channel Tunnel
  d The Golden Gate Bridge

3 Read the text again and answer the questions (1–9) below.

  1 Where is it?
  2 What is it?
  3 How long is it?
  4 Who built it?
  5 How did they build it?
  6 What are TBMs?
  7 How big are TBMs?
  8 How long did it take to build?
  9 When did it open?

4 First, underline the question words in Exercise 3. Then use them to complete these questions.

  1 _____ many Roman roads are there in Europe?
  2 _____ designed St Paul's Cathedral in London?
  3 _____ is the name of the famous bridge in San Francisco?
  4 _____ was the Eiffel Tower built?
  5 _____ is the Corinth Canal?

The … is between Britain and France. It's more than 20 kilometres long. It was built by British and French engineers. They started on opposite sides and met in the middle under the sea. They used specially-designed tunnel boring machines (TBMs) to dig the tunnels through the rock under the seabed. TBMs are enormous machines for digging tunnels. The machines used to dig the main tunnels were about 8.5 metres in diameter and 250 metres long. Work started in 1987 and the teams met under the seabed in 1991. It is a rail tunnel. The first passenger train went through in 1994.

Vocabulary

5 Complete the texts by putting one word in each space. Use the words in the box. Check the meaning of any new words in the glossary or your dictionary.

> across ∎ around ∎ between ∎ over ∎ through ∎ under

The Panama Canal is a 64km waterway (**1**) _____ the Atlantic and Pacific Oceans. Before the canal was opened, ships had to travel thousands of miles (**2**) _____ South America. To build the canal, engineers had to dam a major river, and dig a channel (**3**) _____ a mountain ridge.

Tower Bridge is an openable bascule bridge, designed by Horace Jones in 1886. It goes (**4**) _____ the River Thames in London. Thousands of vehicles drive (**5**) _____ it every day. Tall ships cannot pass (**6**) _____ Tower Bridge, instead, the roadway parts and lifts to let them through.

Writing and Speaking

6 Write questions about a building, tunnel, or dam. Use *who, what, when, where*, and *how*. Make sure you know the answers!

For example
How old is it? Where is it?

7 Work with a partner. Ask and answer the questions you wrote in Exercise 6.

> ▶ ### Get real
> Find out about a major new engineering project. Where is it? What will it do? What problems do the engineers have to solve to build it?

16 | Bright spark

Before you start

1 Write down, in your own language, the type of energy used to:

- provide light
- power an old-fashioned clock
- heat buildings
- power a modern watch
- ride a bicycle

2 Compare your answers with the rest of the class. What are the English words for these types of energy?

Reading

3 Read the text quickly and choose the correct answers the questions below.

1 Trevor Baylis is
   - a a doctor
   - b a TV presenter
   - c an inventor

2 The text is about
   - a radios
   - b a clever idea
   - c Africa

4 Read the text again and decide if the sentences (1–5) below are true (T) or false (F).

1 Trevor Baylis had his idea when he watched a TV programme. T/F
2 He wanted to give people information about AIDS. T/F
3 His radio is powered in two different ways. T/F
4 The idea has been successful. T/F
5 Only radios can have clockwork power. T/F

IN 1991, Trevor Baylis saw a television programme about people in Africa with AIDS. A doctor in the programme said he wanted to give everyone in his country information about the illness but very few people had televisions or radios. The problem was that radios were very expensive because the batteries cost more than a week's food for a family.

Trevor Baylis had a clever idea: a clockwork radio that didn't need batteries. He designed and developed a mechanism where the energy stored in a wound up spring could be used to drive a generator to power the radio. He also added a panel to convert solar energy into electrical energy.

Trevor Baylis' environmentally-friendly radio has won lots of awards. The technology can be used in anything that needs batteries and it is perfect for countries where electrical power is unreliable or very expensive. The wind-up technology is now used in the new generation of Apple e-Mate computers.

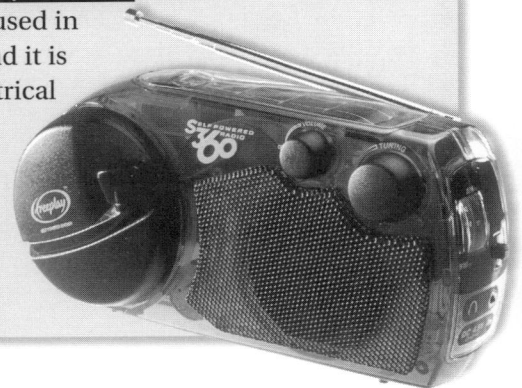

Vocabulary

5 Complete the definitions (1–5) below with the highlighted words in the text. Use the glossary or your dictionary to help you.

1 _____ means good for the health of people and the world.
2 _____ is power produced by a wound up spring.
3 Something that often doesn't work is _____.
4 A _____ converts mechanical power to electrical energy.
5 Power from the sun is _____.

Writing

6 Write a paragraph about another invention. You can use your own idea or these notes about the personal stereo:

- first personal stereo: Sony Walkman
- invented: 1979
- invented by: Sony employees in Japan

Start your paragraph like this:
Before 1979 the only way to listen to music outside was …

> ### ▶ Get real
> Find out about a modern inventor from your country. What did he/she invent and why is their invention important?

17 Servant or master?

Before you start

1 How many senses have you got? What are they? Why are they important?

2 Match the verbs in column A with the parts of the body in column B and the sense nouns in column C.

A verbs	B parts of the body	C sense nouns
see	nose	smell
hear	hands	touch
smell	eyes	sight
taste	ears	hearing
touch	mouth	taste

3 What is the difference between *see* and *look, hear* and *listen*?

Reading

4 Read the text quickly and choose the correct answers to the questions below.

1 Are the paragraphs about
  a lots of topics?
  b one topic?

2 Which is the best title?
  a Imaginary robots in film and fiction
  b Robots: the fantasy and the facts

A We can thank the world of literature for the words *robot* and *robotics*. The word *robot* was first used by the Czech playwright Karel Capek in his 1921 play, *RUR* (*Rossum's Universal Robots*). **(1)** _____. Asimov used it in a short story in 1941.

B Robots often star in films too, for example dangerous machines like Terminator or cute ones like R2D2 in *Star Wars*. **(2)** _____ Industrial robots don't have personalities and they don't think like people. Most real robots are designed to save people from dangerous jobs **(3)** _____ or boring, routine work **(4)** _____.

C A simple robot is made of:
- A mechanical device **(5)** _____ that can react to its environment.
- Sensors that **(6)** _____ give information to the device.
- Systems or computer programs that **(7)** _____ give the device instructions.

5 Read the text again and put the sentences and phrases (a–g) below in the correct places (1–7).

a in factories, laboratories, or warehouses
b In this play machines behave like people
c like an arm
d can 'see' the environment and
e like handling nuclear or radioactive materials
f The reality is less exciting.
g understand the messages from the sensors and

6 What is the connection between human senses and robots?

Vocabulary

7 Find words in the text that mean:
1 always done in the same way
2 respond to a change
3 a piece of equipment designed to do a particular job
4 part of a machine that can sense heat, light, etc.

Writing and Speaking

8 Design a robot to do a dangerous or boring job for you. Draw a rough sketch and make notes about how it works.

For example
What the robot is for — *It is to pick up the socks in my bedroom.*
How it works — *The sensor smells ..., the arm ....*

9 Discuss your ideas with a partner. Comment on your partner's idea.

For example
What powers the robot? Where does it put the socks?

▶ **Get real**
Find examples from your own country of robots in a story, a film, and in real life. Tell your class about them. Who found the most interesting or technologically advanced robots?

18 | Gadgets

Before you start

1 Look at the pictures (1–4). What do you think the things are? Discuss your ideas with the rest of the class.

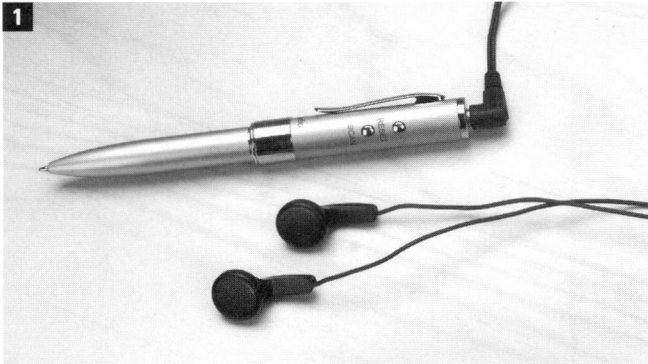

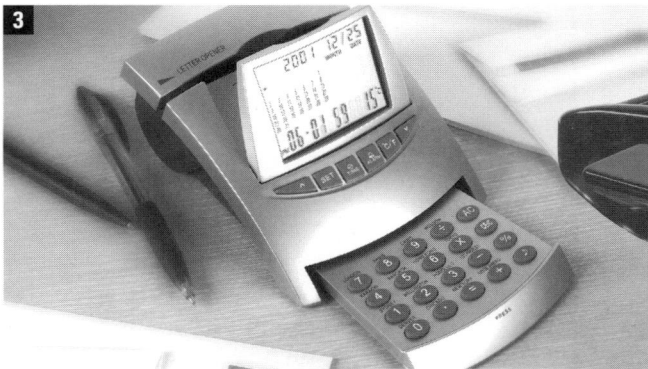

Reading

2 Read the descriptions (A–D) and match them with the pictures (1–4).

A LETTER OPENER CLOCK £19.99

Desktop clock, thermometer, calendar, and letter opener

This gadget has got lots of helpful information – with the added benefit of an automatic electric letter opener. The LCD display shows date, time, and temperature in °C or °F. In addition there is an alarm clock, a calculator, and the times in 15 cities around the world.

Letter opener uses 2 x AA batteries (not supplied). 5.5 x 12 x 9.5cm.

C FEET WASHER £19.99

The best thing for your feet

Designed for shower or bath, this vinyl mat cleans and massages your feet – and you don't need to bend down or stand on one leg! Suction cups hold it safely while you stand on the 1,500 relaxing 'fingers'. 2.5 x 14.5 x 27.5cm.

B RADIO PEN £14.99

Sounds as good as it writes

This pen looks beautiful, feels great to write with – but it sounds better in the ears. It's got a secret radio in the top! Wear the earphones and enjoy music while you work.

Button batteries included. 14cm long.

D BED GLASSES £29.99

How to read or watch TV – flat on your back

These glasses are perfect for sick people who must stay in bed, or for people who like to relax with a book or watch TV while lying flat on the floor or sofa. The plastic frame contains two glass prisms that deflect your vision by 90°. The lenses are first-class and you can wear them over your normal glasses.

3 Read the texts (A–D) again and match the sentences (1–6) below with the gadgets.

1 These two don't need batteries.
2 This does two things.
3 You use this standing up.
4 You use these lying down.
5 This can tell you how hot it is.
6 You get free batteries with this.

Vocabulary

4 Complete the definitions (1–7) with the ▮highlighted▮ words in the text. Use the glossary or your dictionary to help you.

1 A _____ is a good thing.
2 A _____ is a triangular block of glass.
3 _____ are the pieces of glass you look through.
4 _____ holds/attaches with air pressure.
5 _____ is a type of plastic.
6 _____ means hidden.
7 To _____ means to change direction.

Speaking

5 How useful are the gadgets? Put them in order (1 = most useful, 4 = least useful). Discuss your ideas with a partner and agree on an order. Explain your list to the rest of the class. Do other students agree with you?

6 Think of a gadget that you use in your home, for example, a TV remote control, a potato peeler. (You don't need to know the English word). Plan how to describe it, for example:

Where is it used?
Who uses it?
Why is it useful?
What is it made from?
How much does it cost?

7 Without naming the gadget, describe it to your class. Can they guess the gadget? Does anyone know the English word?

For example
You use this in the kitchen. You use it when you're cooking potatoes or carrots. It's easy to use and safer than a knife. It's made from metal. It doesn't need batteries. It's not expensive.

Writing

8 What information is included about each gadget? Add to this list.

• The name of the gadget.
• What it does.
• …

9 Design a gadget. Choose one of the gadgets below or your own idea. Do a rough drawing of your gadget and write draft information about it. Use your list from Exercise 6 to help you.

• A gadget to cut your toenails without bending over.
• A gadget to exercise your dog without going outside.
• A gadget to keep your younger brother/sister out of your bedroom.
• A gadget to clean your shoes

10 Exchange your work with another student. Can you help each other to improve the drawings and descriptions?

11 Make a class poster of your final drawings

> ### ▶ Get real
> Use the Internet or magazines to find examples of strange or unusual gadgets. Bring them into class. Decide who has found the strangest or most unusual gadget.

Before you start

1 Do you like high places or are you afraid of heights (vertigo)? Choose two words in your own language to describe standing on a tall building. Compare your words with the rest of the class. How do you say these words in English?

Reading

2 Read the text quickly and choose the correct answers to the questions below.

1 Where is the text from?
   a An engineering book about bridges
   b A tourist guidebook

2 Which is the best title for the text?
   a The Clifton Suspension Bridge
   b Isambard Kingdom Brunel

Vocabulary

3 Complete the text by putting one word in each space. Use the words in the box. Check the meaning of any new words in the glossary or your dictionary.

> aeroplane ■ built ■ computer ■ designed ■
> engineers ■ fixed ■ parachute ■ perfect ■ pilot ■
> vehicles

Reading

4 Read the text again and decide if the sentences (1–5) below are true (T) or false (F).

1 The Clifton Suspension Bridge is a
  moveable bridge.   T/F
2 It was designed and built in the
  twentieth century.   T/F
3 The designer was a famous British
  engineer.   T/F
4 The design of the bridge is very good.   T/F
5 Pilots fly under the bridge every day.   T/F

5 Look at the diagram and the dimensions. Write the dimensions in the correct places.

Dimensions
span   214m
height above river   75m
height of towers   26m

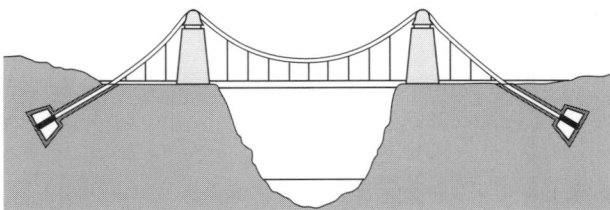

The Clifton Suspension Bridge is a
(1) _____ bridge which means it doesn't open or move to allow boats through. It was (2) _____ in the 1830s by one of Britain's greatest nineteenth-century (3) _____, Isambard Kingdom Brunel. The bridge was actually (4) _____ in the 1860s. (5) _____ analysis of the design shows that many of the ideas are almost (6) _____. When the bridge opened it was for carriages pulled by horses but 150 years later it carries 12,000 cars and lorries a day, that's over four million (7) _____ a year.

> **Strange but true**
>
> **1** In 1885 a young woman jumped off the Clifton Suspension Bridge. Her large, fashionable, nineteenth-century skirt acted as a (**8**) _____ and she landed safely after a 75m fall. She lived into her seventies.
>
> **2** The first (**9**) _____ flew under the bridge in 1911. The last plane was a jet travelling at 720kph in 1957; the (**10**) _____ crashed the plane and died.

▶ Get real
Find out about a famous bridge or building in your country. Who designed it and when was it built? Draw a diagram and label the dimensions.

20 Bridge disaster

Before you start

1 Have you ever done either of the things below? Could you have avoided it? How? Tell your class what happened and what could have prevented it.

- lost something important?
- injured yourself in an accident?

Vocabulary

2 Look at these words from the text. Check the meaning of any new words in the glossary or your dictionary.

> bridge ▪ collapse ▪ cracked ▪ disaster ▪ enquiry ▪ killed ▪ substandard

Reading

3 What do you think the text will be about? Discuss your ideas with the rest of the class. Read the text and check.

THE TAY BRIDGE in Scotland was designed and built by Sir Thomas Bouch in the nineteenth century. The bridge, which was over 3km long, opened in 1878 and fell down in a winter storm in 1879. A train carrying 70 people was on the bridge at the time and all the people were killed.

There was an enquiry into the Tay Bridge disaster to find out why the accident happened. One of the conclusions was that the design and construction were based on how quickly and cheaply it could be built; safety and strength were not thought about properly. Another conclusion was that it was a very cold winter and the iron may have cracked when it contracted. Also, the design was based on experience rather than the more scientific and accurate calculations used today.

During the twentieth century engineers used computers to do a detailed structural analysis of the design used for the Tay Bridge. The results confirm that the design of the bridge was definitely substandard.

4 Read how the Tay Bridge collapsed. Match the sentences (1–5) with the diagrams (a–e) below.

1 There was strong wind when a train was crossing the bridge.
2 The base of one of the columns lifted.
3 This (lifting) caused the strengthening parts of the structure to collapse.
4 The whole structure started to fall.
5 While the structure was falling, the girders collapsed in the opposite direction, causing the pier to collapse completely.

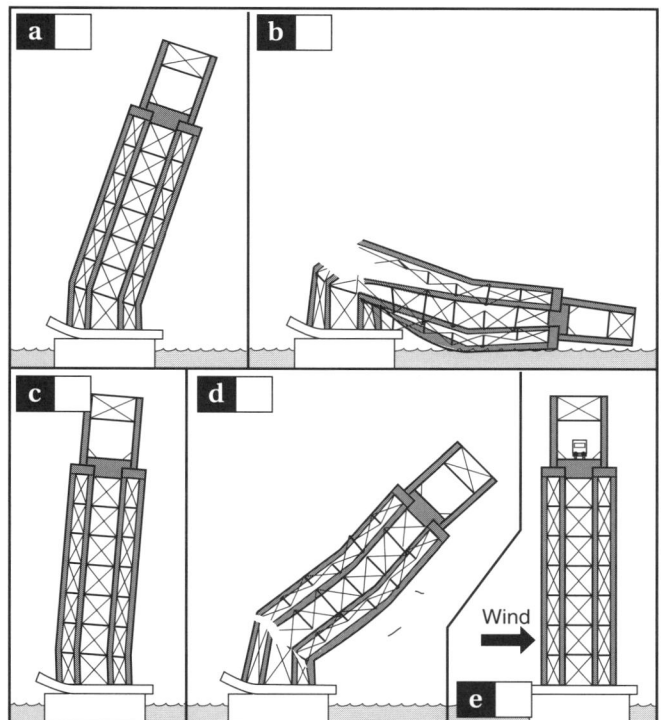

Speaking and Writing

5 Work with a partner. Imagine you are going to design a building, a bridge, or another structure. Make a list of the questions you should ask before you start.

Example:
Designing a school
- *how many children are going to attend the school?*
- *how many classrooms will they need?*

6 Compare your ideas with other students. Can you help them to add to their list?

▶ ## Get real

Find out about another engineering or design disaster. Which questions should the designer have asked before he/she started work?

Before you start

1 Look at the pictures. Do you know what and where these structures are? Are there any ancient structures in your country?

Reading

2 Read the text and choose the best title, A, B, or C.

A Good jobs and good pay
B Ancient engineering
C The rulers of China and Egypt

Title: _____

THE GREAT WALL OF CHINA was built across northern China to protect the **population**. Before the third century BC there were lots of smaller walls and these were joined together to make one long, **defensive** wall. The work was done by enormous gangs of forced **labourers** and many of them died doing the work. The wall is over 2000km long, 3.5m high, and 4.5m wide at the top. It is made of earth covered with stone.

THE EGYPTIAN PYRAMIDS are a famous symbol of ancient Egypt. The stone structures were usually **tombs** for pharaohs. The pyramids have square bases with sloping sides which meet at an **apex**. The first pyramid was built in about 2600 BC and is over 140m high. One of the biggest pyramids is made of enormous stone blocks which weigh up to 200 tonnes each. It is estimated that 20–25,000 people worked for 20 years to build each pyramid.

3 Use the information in the text to answer the questions (1–6) below about 1) The Great Wall of China and 2) the first pyramid in Egypt.

1 Where is it?
2 What is it?
3 When was it built?
4 What is it made of?
5 Who built it?
6 How big is it?

4 Read the text again and decide if the sentences (1–5) below are true (T) or false (F).

1 The Great Wall of China was to keep people safe. T/F
2 Building the Great Wall was easy for the workers. T/F
3 The pyramids were built before the Great Wall of China. T/F
4 Pyramids are lots of different shapes. T/F
5 The Egyptian pyramids were built protect people. T/F

Vocabulary

5 Match the **highlighted** words in the text with the definitions (1–5) below.

1 A place where people are buried.
2 The top, or highest part of something.
3 Protecting somebody against attack.
4 People who do hard physical work outdoors.
5 All the people who live in a country.

Writing

6 Use the notes below to write a paragraph about Hadrian's Wall (the northern boundary of the Roman Empire).

- Between England and Scotland
- A defensive wall
- Built in 122–6 AD
- Made of earth and stone
- Built by soldiers
- 117km long, 6.5m high, 3m wide

▶ **Get real**

Find out about an ancient structure in your country. Answer the questions in Exercise 3 about it.

Before you start

1 What does this Chinese proverb about learning mean? Do you agree with it? How do you like to learn?

I hear and I forget,
I see and I remember,
I do and I understand.

Vocabulary

2 Match the words in the box with labels (a–j) on the diagrams below.

> bottom ■ centre ■ corner ■ diagonal ■ fold ■
> horizontal ■ point ■ side ■ top ■ vertical

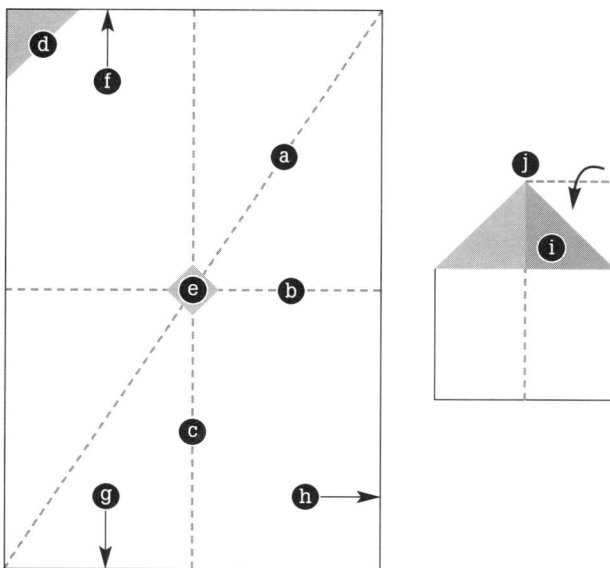

Reading

3 Read the origami instructions (1–9) and match them with the diagrams (a–i).

4 First, follow the instructions and make a paper plane. Then throw it in the air at an angle of 45 degrees. Did it fly?

Writing

5 Work out how to make a simple paper object. Use your own idea or choose one of these:

> a cube ■ an envelope ■ a boat

Make the object, write draft instructions, and draw rough diagrams for each step.
Ask your partner to read your draft and follow the diagrams.
If they can't make the object, improve your instructions and diagrams.
Ask another person to try to make the object.

Origami: A paper plane

1. Take a piece of A4 paper (210x297mm).

2. Make a vertical fold down the centre and open it out.

3. Fold the top two corners to the centre line to form two diagonal lines.

4. Fold again so the diagonal lines meet on the centre line.

5. Fold a horizontal line across the middle and bring the point to the middle of the bottom edge.

6. Fold two diagonal lines so the two short edges at the top meet on the centre fold.

7. Lift the point and make a horizontal fold. The nose of the plane meets the point at the top.

8. Fold along the centre fold.

9. Make two folds as shown to make the wings.

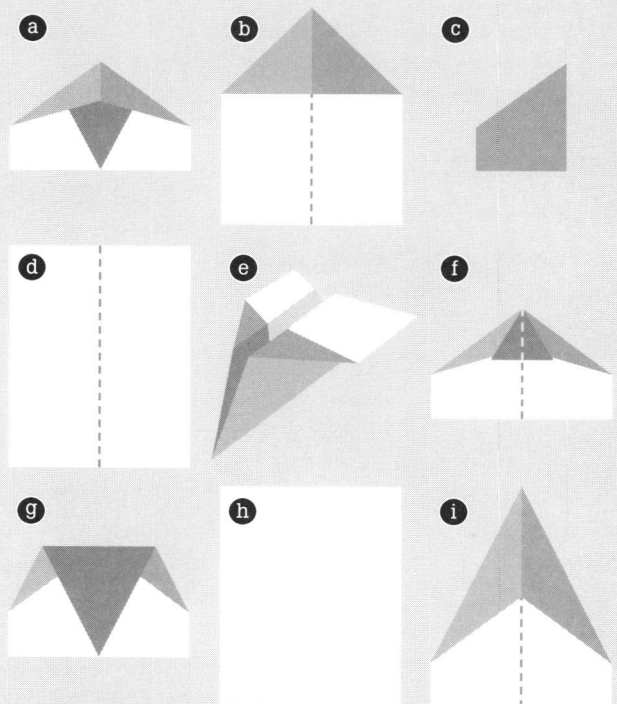

▶ *Get real*

Find examples of different types of instructions, for example furniture assembly instructions with diagrams, or cooking recipes. Which style do you find easiest to understand?

Before you start

1 What do these abbreviations stand for? Match the abbreviations in the box with the full forms (1–9) below.

cm ■ kg ■ l ■ ml ■ g ■ $(x)^2$ ■ $(x)^3$ ■ km ■ m

1 centimetre /'sentɪmi:tə/
2 gram /græm/
3 kilogram /'kɪləgræm/
4 kilometre /'kɪləmi:tə/
5 litre /'li:tə/
6 metre /'mi:tə/
7 millilitre /'mɪlɪli:tə/
8 cubic /'kju:bik/
9 square /skweə/

2 Are the words the same in your language? Why is it useful to have standard international systems?

3 How do you pronounce the words in your language? Look at the pronunciation of the words in English and say them.

Vocabulary

4 What are the measurements in Exercise 1 used for? Complete the sentences (1–8) below by putting one word in each space. Use the words in the box.

area ■ capacity ■ distance ■ length ■ liquid (quantity) ■ speed ■ weight ■ height

Did you know?

1 The _____ of the Eiffel Tower in Paris is about three hundred metres.
2 The _____ of the Charles Bridge in Prague is five hundred and sixteen metres.
3 The surface _____ of Lake Balaton in Hungary is five hundred and ninety-three square kilometres.
4 The maximum _____ limit on expressways in Poland is one hundred and ten kilometres per hour.
5 The _____ of the bell in Dubrovnik's city tower is two thousand kilograms.
6 The _____ between Bratislava and Budapest is about two hundred kilometres.
7 A magnum champagne bottle can hold one point five litres of _____.
8 The engine _____ of a Formula One car is three thousand cubic centimetres.

Note: In English, we say 5m x 7m as *five metres by seven (metres)* when we are talking about area. In mathematics, 5x7 is *five times seven* or *five multiplied by seven*.

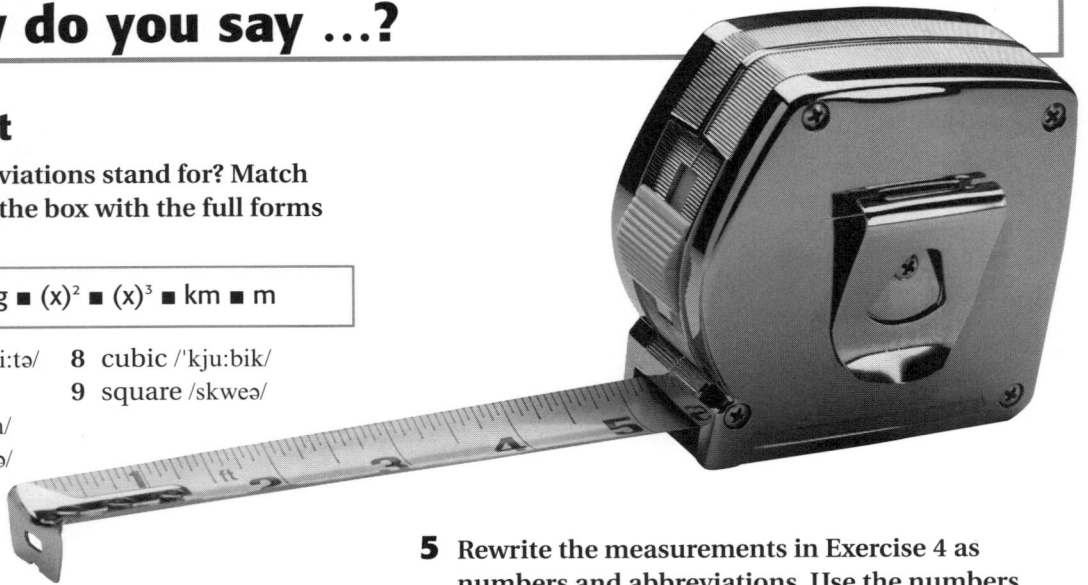

5 Rewrite the measurements in Exercise 4 as numbers and abbreviations. Use the numbers and abbreviations in the box.

516m ■ 110kph ■ 3000cc (or cm^3) ■ 200km ■ 300m ■ 1.5l ■ $593km^2$ ■ 2000kg

6 Rewrite the measurements (1–9) below as numbers and abbreviations.

1 twenty-two kilometres per hour
2 two litres
3 one point five square metres
4 six square kilometres
5 fifty millilitres
6 eighteen kilograms
7 one hundred and thirty grams
8 one point five metres by fifty centimetres
9 nought point seven five cubic metres

Writing and Speaking

7 Write true answers to these questions. Use words, not numbers.

1 What area is your classroom?
2 How tall are you?
3 What is the speed limit on the roads in your town?
4 How fast can you run?
5 What is the area of your desk?
6 How much does your school bag weigh?
7 How much did you weigh when you were born?
8 How far is it from your town to the capital city?

8 Choose six answers from Exercise 7 and read them to your partner. Can he/she identify the question?

▶ ### Get real

Find out about the history of measurement. For example, what are the connections between human bodies and measurement? Which countries developed the earliest standard systems? Why did they need them?

Before you start

1 What are these things? What have the words got in common?

> Biro ■ Braille ■ guillotine ■ Hoover ■ Jacuzzi ■ Levis ■ Stetson

Reading

2 Put these standard international (SI) units into the correct column.

> amp ■ Celsius ■ curie ■ hertz ■ joule ■ kelvin ■ newton ■ ohm ■ pascal ■ volt ■ watt

Chemistry (1)	Electricity (6)	Physics (2)	Temperature (2)

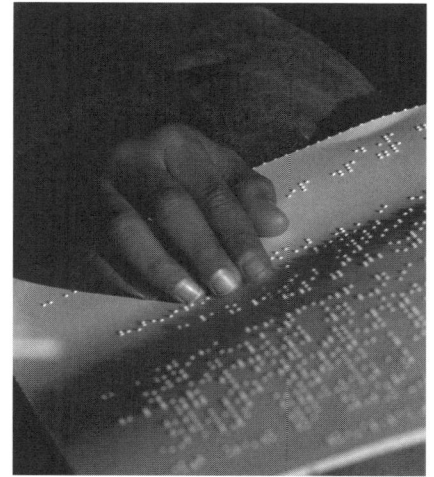

A blind person reading Braille text

3 Now complete the definitions (1–11) below with the units from Exercise 2 and the people in the box.

> André Marie Ampère (1775–1836) ■ Anders Celsius (1701–1744) ■ Marie Curie (1867–1934) ■ Heinrich Hertz (1857–1894) ■ James Prescott Joule (1818–1889) ■ Lord Kelvin (1824–1907) ■ Georg Simon Ohm (1787–1854) ■ Blaise Pascal (1623–1662) ■ Sir Isaac Newton (1643–1727) ■ Count Alessandro Volta (1745–1827) ■ James Watt (1736–1819)

1 A _____ is a unit of pressure equal to one newton per square metre. It's named after _____ a French scientist.

2 A _____ is a unit of force. It's named after _____ an English mathematician.

3 _____ is the temperature scale that has the freezing point of water as 0° C and the boiling point as 100° C. The scale was developed by a Swedish astronomer, _____.

4 A _____ is an amount of electric power. It is equal to one joule per second. It's named after _____, a Scottish engineer and inventor.

5 A _____ is a unit of electric force. It's named after _____, an Italian physicist and pioneer in the study of electricity.

6 An _____ is a unit of electric current. It's named after _____ a French mathematician and physicist, a pioneer in electrodynamics.

7 An _____ is a unit of electrical resistance named after _____ a German physicist.

8 A _____ is a unit of energy named after _____ a British physicist.

9 _____ is the temperature scale that registers absolute zero (−273.15 C) as 0°K. It's named after _____ a British scientist.

10 A _____ is a frequency equal to one cycle per second. It's named after _____ a German physicist.

11 A _____ is a unit of radioactivity. It's named after _____ a Polish-born chemist who discovered radioactivity in several elements.

Vocabulary

4 Read the definitions in Exercise 3 again. Find words that mean:

1 studies the elements and their compounds
2 studies the universe
3 studies the physical properties of materials
4 thinks of new machines
5 develops new ideas about a subject

Speaking

5 Discuss these questions with a partner:

- Which of the words in Exercise 2 and 3 are you familiar with in your language?
- Is anything named after a person in your country?

> ### ▶ Get real
> Find out which things in this list are named after people. Can you add similar words from your language?
> - Mouse (for a computer)
> - Bunsen (burner)
> - Diesel
> - Geiger (counter)
> - Laboratory
> - Morse (code)
> - Tarmac
> - Text (book)

Before you start

1 Match the Arabic and Roman numbers.

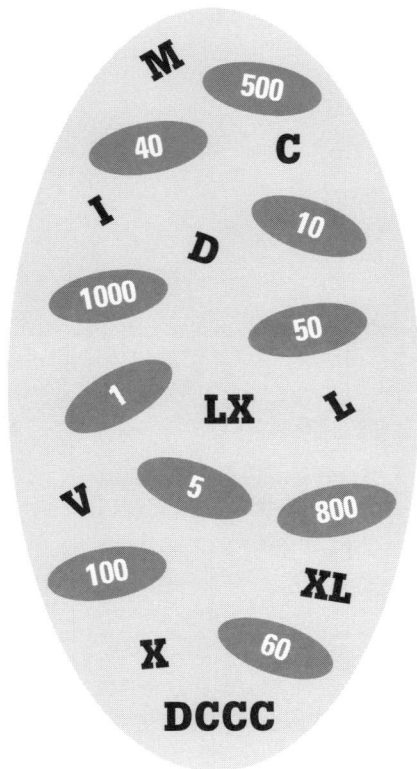

M · 500 · 40 · C · I · 10 · D · 1000 · 50 · 1 · LX · L · V · 5 · 800 · 100 · XL · X · 60 · DCCC

2 Why are Arabic numbers used in mathematics?

Vocabulary

3 Match the numbers in the box with the words (1–9) below.

$\frac{1}{2}$ · 1,000,000/1m · 2.5 · 327 · 2,580 · 0 · $\frac{1}{4}$ · $\frac{2}{3}$ · 3.6%

1 two thirds
2 three point six per cent
3 a quarter
4 zero/nought
5 two point five
6 one million
7 two thousand, five hundred and eighty
8 three hundred and twenty-seven
9 a half

Note: In English, you write a point (.) not a comma (,) in decimal numbers. You say the numbers after the point separately, for example 23.34 is 'twenty-three point three four'.

Reading

4 Complete the text by putting a word or number in each space (1–10). Use the words and numbers in the box.

half · -5° · -40° · 88% · Hundreds · 200 · 14,000 · 1989 · 4000 · 5000m²

@ Icehotel

Address: @ ice_hotel

JUKKASJARVI
ICEHOTEL

THE JUKKASJARVI ICEHOTEL in Sweden is an interesting and cold place for a holiday. It started life as an igloo (a small house made of snow) at an art exhibition in (**1**) _____.

(**2**) _____ of people visited the exhibition and some even slept there, so the builders decided to make it a hotel.

The Icehotel is open for less than (**3**) _____ of the year. Every May it melts and every November it is rebuilt. It now measures (**4**) _____ and it needs (**5**) _____ tons of ice and 30,000 tons of snow to build it. This actually means that it is more than (**6**) _____ snow.

The temperature inside the hotel is usually about (**7**) _____. Outside in Jukkasjarvi itself the temperature can be much lower even as low as (**8**) _____!

Last year more than (**9**) _____ visitors travelled (**10**) _____ km north of the Arctic Circle to sleep in thermal sleeping bags. They got a cool reception!

Internet zone

Speaking

5 Work in pairs, Student A and Student B. Dictate your numbers to your partner. Can your partner write them correctly?

Student A
• thirty-four point five percent
• six point nine seven
• one third
• four thousand, five hundred and sixty-seven

Student B
• three thousand, nine hundred and fifty-eight
• fifty-five percent
• a half
• seven point six five

▶ Get real

Find out the answers to these questions:
• Who introduced Arabic numbers to European maths?
• Who developed the idea of 'zero'?

26 | It's all just numbers

Before you start

1 Do you know how many people live in your town and in your country? Where can you find this information?

Reading

2 Read the text quickly and choose the correct answers to the questions (1–3) below.

1 Where is the text from?
   a A government information leaflet
   b A teenage magazine

2 What is the text about?
   a how many children watch TV in Britain
   b how many children there are in Britain

3 Who is this information useful for?
   a people planning educational resources
   b teachers and parents

P⊕PULATION

There were **(1)** twelve point one million children aged under **(2)** sixteen in **(3)** two thousand: **(4)** six point two million boys and **(5)** five point nine million girls. This is fewer than in **(6)** nineteen seventy-one, when there were **(7)** fourteen point three million children. In **(8)** two thousand, **(9)** thirty per cent of children in the UK were under five, **(10)** thirty-two per cent were aged five to nine years and **(11)** thirty-eight per cent were aged ten to fifteen. These proportions were similar in the **(12)** nineteen seventies.

3 Write the numbers in *italics* in figures.

4 Read the text again and decide if the sentences (1–3) below are true (T) or false (F).

1 There are more boys than girls in Britain. T/F
2 The total number of children has increased since a census in 1971. T/F
3 In 1971 the same percentage of children were under five. T/F

5 Use the information in the text in Exercise 3 to label the bar chart and pie chart.

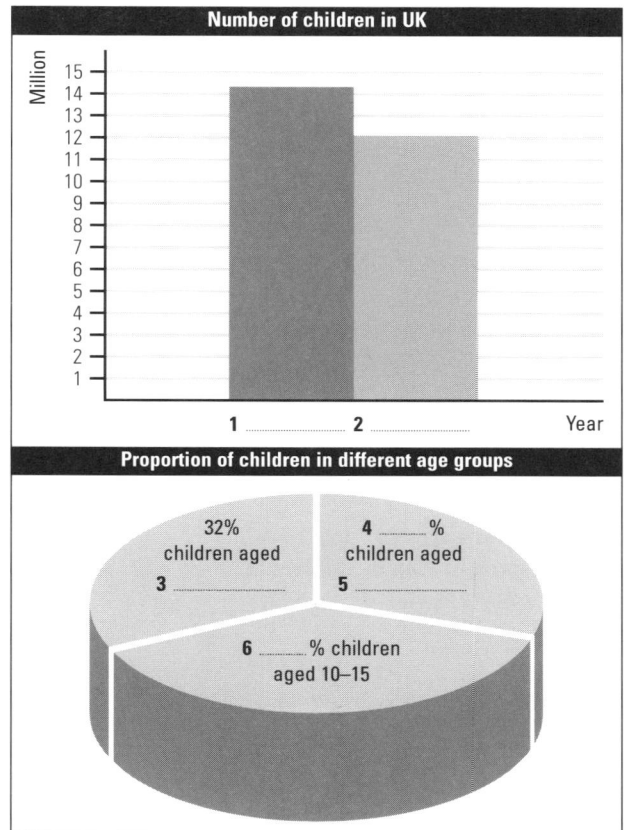

Number of children in UK

Million
15
14
13
12
11
10
9
8
7
6
5
4
3
2
1

1 _____ 2 _____ Year

Proportion of children in different age groups

32% children aged 3 _____

4 _____ % children aged 5 _____

6 _____ % children aged 10–15

Writing and Speaking

6 Do a survey in your class. First, write your question/s and decide how you will record the answers. Choose one of the topics below, or your own topic.
- how much people weigh
- your school canteen
- amount of time spent watching TV every day

7 Ask the other people in your class your questions and write down their answers.

8 Transfer the figures to a pie chart and write a paragraph describing your findings.

For example:
Twenty people were asked, 'What do you think of the school canteen?' Ten percent are happy with the school canteen, 5% didn't know there is a canteen, 50% prefer to bring sandwiches, and the other 35% don't feel very well.

▶ *Get real*
Use reference books or the Internet to find some unusual, strange, or interesting statistics. Bring them into class and create a *Did you know?* board.

Before you start

1 Have you ever had a part-time or work experience job? Tell your class:

- what your job was
- how you got it

Reading

2 Read the curriculum vitae (CV) quickly and choose the correct answers to the questions below.

1 What is a CV?
  - **a** A description of someone's family, education, likes and dislikes.
  - **b** A description of someone's education, work experience, and skills.

2 How is a CV arranged?
  - **a** under headings
  - **b** like a letter

3 Read the CV again and decide if the sentences (1–7) below are true (T) or false (F).

1 Gavin Alvarez lives in Cambridge.   T/F
2 He is a student at Cam College.   T/F
3 He passed his GCSEs in 2001.   T/F
4 He has had Saturday and holiday jobs since 2000.   T/F
5 He left Cam College in 2000.   T/F
6 He is quite good at languages.   T/F
7 He isn't interested in technology.   T/F

NAME	Gavin H Alvarez
ADDRESS	26 Dryfield Road Cambridge CB2 2DS
TELEPHONE NUMBER	01223 3268452
E-MAIL ADDRESS	gavinhalvarez@btinternet.com
DATE OF BIRTH	14 June 1984

EDUCATION

1995–2000	Graves High School for Boys Graves Avenue Cambridge CB3 4RG
2000–2002	Cam College of Engineering and Technology Birch Road Cambridge CB6 7YT

QUALIFICATIONS

2000	GCSEs: English, Maths, General Science, Design and Technology, French, Spanish, Art, and History
2001	Level 1 Engineering and Technology foundation course
2002	Level 2 Computing course specializing in software development

WORK EXPERIENCE

AUGUST – SEPTEMBER 2000	Temporary job as IT assistant at Norris's Aeronautics, Cambridge.
OCTOBER 2000 – JUNE 2002	Saturday and holiday job testing computer games at Silicompany, Cambridge.
OTHER INFORMATION	bi-lingual in Spanish and English; clean driving licence
INTERESTS	developing computer games, member of college football team, photography, and playing the guitar
REFEREE	Ms Daisy Valentine (course tutor) Cam College of Engineering and Technology Birch Road Cambridge CB6 7YT

4 Read these two advertisements. Which job is best for Gavin?

a

GAMES4U

wants young, dynamic people to develop the next generation of computer software.
Foreign languages an advantage.
Send CV and covering letter to …

b

BOOKS FOR STUDENTS

need two people aged 18–20 to work in their engineering and technology department.

To apply, send CV and covering letter to …

Vocabulary

5 Read these phrases from the CV and the advertisements. Choose the correct meaning (a or b) of the words in *italics*.

1 … a *covering* letter ….
   a a letter to give more information
   b a letter to hide a CV

2 foreign languages an *advantage* …
   a it will help if you can speak a foreign language
   b it is essential that you can speak a foreign language

3 a *clean* driving licence
   a your licence isn't dirty
   b you haven't done anything illegal in a car

6 Complete the table with words from the CV and the advertisements.

noun/person	verb	noun/subject
developer	1	2
specialist	3	
tester	*to test*	
engineer		4
technician		5

Writing

7 Write your own CV in English using qualifications you already have, or ones that you think you might get in the future. Use Gavin's CV as a model for your writing.

Name	
Address	
Telephone number	
e-mail address	
Date of birth	
Education	
Qualifications	
Work experience	
Other information	
Interests	
Referee	

▶ ***Get real***
Are the rules for CVs the same in your language as in English? Are there any differences?
Look at advertisements for jobs that interest you. What do you have to do to apply?

Before you start

1 Do you ever write these types of text? Which ones do you write most often? Who do you write them to?

- postcards?
- e-mails?
- letters?
- text messages?

Reading

2 Read the letters (A and B) and choose the correct answers to questions 1–5 below.

1 Letter A is *to/from* Katy Evans.
2 *Mrs Lee/Katy Evans* works at the college.
3 Katy Evans *is/wants to be* a student at the college.
4 Katy Evans wants information about *a basic/an advanced* course.
5 Mrs Evans sends *exactly what/more than* Katy asked for.

3 Identify the features (1–6) on each letter. Use the features (A–F) below.

A The receiver's address
B The receiver's name
C The sender's address
D The sender's phone number
E The sender's name
F The date

4 Explain why Katy Evans wrote *Dear Sir/Madam*.

A

1 28 High Avenue
Harlow
Essex
CM16 7AY

2 01279 5743461

3 Admissions Secretary
Stevenage Technical College
West Road
Stevenage
Herts ST6 8PI

4 30 April 2003

5 Dear Sir/Madam,
Please send me details of your foundation course in computer engineering.

Yours faithfully,

6 Katy Evans

B

1 Stevenage Technical College
West Road
Stevenage
Herts
ST6 8PI

2 01438 7546392

3 Ms K Evans
28 High Avenue
Harlow
Essex
CM16 7AY

4 17 May 2003

5 Dear Ms Evans,

Thank you for your enquiry. I enclose the details that you requested and some additional information about the college's facilities.

Please contact me if you require any further information.

Yours sincerely,

6 Mrs LH Lee
Admissions Secretary

Vocabulary

5 Read the messages (a–l). Are they formal (F) or informal (I) ?

a Please contact me to arrange another appointment.

b I'll show you round ...

c Call me and we'll fix a time to meet.

d I regret to inform you that your application has not been successful.

e Please supply additional information about your experience.

f ☎ cu@10

g Tell me more about what you did.

h I'd like to introduce you to ...

i There will be an opportunity to visit ...

j Bad luck about the job.

k Your interview has been arranged for ten o'clock.

l This is ...

6 Match the formal and informal messages with the same general meaning.

Speaking

7 In your own language, discuss the differences between formal and informal English. Are the differences the same in your language?

Writing

8 Rewrite these formal phrases in informal English.

We'd be delighted if you could join us for drinks.
Let's have a drink.
1 In response to your recent enquiry ...
2 The receptionist will direct you to the correct room.
3 I look forward to hearing from you.
4 I hope that will be acceptable.

9 Draft a formal letter requesting information. You can use your own idea or one of these:

- information about a cycling holiday. Write to: Mr B Roberts, Cycling Heaven, 23 Green Lane, Oxford, OX2 4RD
- information about a course at your local college

10 Exchange your draft letter with a partner. Can you improve your letters? Write the final version of your letter.

▶ ### Get real
Find examples of formal letters in your own language. Compare them with English letters. In which ways are they the same or different?

Glossary

Short forms

[C]	countable	*adv*	adverb
[U]	uncountable	*prep*	preposition
[pl]	plural	*sb*	somebody
adj	adjective	*sth*	something

A

absolute zero /ˌæbsəluːt ˈzɪərəʊ/ *noun* [U] the lowest temperature possible (-273.15°C)

accurate /ˈækjərət/ *adj* exact and correct

across /əˈkrɒs/ *adv, prep* from one side of sth to the other

adjustable /əˈdʒʌstəbl/ *adj* that can be moved into different shapes or positions

advantage /ədˈvɑːntɪdʒ/ *noun* 1 [C] something that may help you to do better than other people 2 [C,U] something that helps you or that will bring you a good result

aerial /ˈeəriəl/ *noun* [C] a long metal stick that receives radio or television signals

aeronautical /ˌeərəˈnɔːtɪkl/ *adj* connected with aircraft

aeroplane /ˈeərəpleɪn/ *noun* [C] a vehicle with wings and engines that can fly

alloy /ˈælɔɪ/ *noun* [C, U] a metal formed by mixing two metals together, or by mixing metal with another substance

aluminium /ˌæləˈmɪniəm/ *noun* [U] a light silver-coloured metal

amp /æmp/ *noun* [C] (abbr **A**) a unit for measuring electrical current

analysis /əˈnæləsɪs/ *noun* [C,U] careful examination

antenna /ænˈtenə/ *noun* (US) = AERIAL

apex /ˈeɪpeks/ *noun* [usually sing] the top or highest part of sth

applicant /ˈæplɪkənt/ *noun* [C] a person who makes a formal request for sth, (**applies for sth**), especially a job, a place at a college, university, etc. ➤ **application** /ˌæplɪˈkeɪʃn/ *noun* [C,U] ➤ **apply** /əˈplaɪ/ *verb*

appointment /əˈpɔɪntmənt/ *noun* [C,U] an arrangement to see sb at a particular time

apprentice /əˈprentɪs/ *noun* [C] a person who works for low pay in order to learn the skills needed in a particular job

architect /ˈɑːkɪtekt/ *noun* [C] a person who designs buildings

architecture /ˈɑːkɪtektʃə/ *noun* [U] 1 the study of designing and making buildings 2 the style or design of buildings

area /ˈeəriə/ *noun* [C,U] the size of a surface

around /əˈraʊnd/ *adv, prep* on all sides; forming a circle

asphalt /ˈæsfælt/ *noun* [U] a thick black substance that is used for making the surface of roads

assembly line /əˈsembli laɪn/ *noun* [C] a line of people and machines in a factory that fit the parts of sth together in a fixed order

automatic /ˌɔːtəˈmætɪk/ *adj* (used about a machine) that can work by itself without direct human control ➤ **automatically** /-kli/ *adv*

automobile /ˈɔːtəməbiːl/ *adj* connected with cars and the manufacture of cars

B

bandwidth /ˈbændwɪdθ/ *noun* [C,U] a measurement of the amount of information that a particular computer network or Internet connection can send in a particular time

bar chart /ˈbɑː tʃɑːt/ *noun* [C] a diagram that uses narrow bands of different heights to show different amounts

bascule bridge /ˌbæskjuːl ˈbrɪdʒ/ *noun* [C] a type of bridge with a section that can be lifted up

base sth on sth *phrasal verb* to form or develop sth from a particular starting point or source

battery /ˈbætri/ *noun* [C] a device which produces electricity

benefit /ˈbenɪfɪt/ *noun* [C,U] an advantage or useful effect that sth has

beware /bɪˈweə/ *verb* **beware (of sb/sth)** (used for giving a warning) (be) careful

Biro™ /ˈbaɪrəʊ/ *noun* [C] a type of pen

boiling point /ˈbɔɪlɪŋ pɔɪnt/ *noun* [C] the temperature at which liquid starts to boil

bore /bɔː/ *verb* to make a long deep hole in sth with a tool

boring /ˈbɔːrɪŋ/ *adj* not at all interesting; dull

bottom /ˈbɒtəm/ *noun* [C, usually sing] the lowest part of sth

brass /brɑːs/ *noun* [U] a hard yellow metal that is a mixture of copper and zinc

break /breɪk/ *verb* to separate into pieces

bridge /brɪdʒ/ *noun* [C] a structure that carries a road or railway across a river, valley, road or railway

build /bɪld/ *verb* to make sth by putting pieces, materials, etc. together

C

cable /ˈkeɪbl/ *noun* [C,U] a set of wires covered with plastic, etc. for carrying electricity or signals

cabling /ˈkeɪblɪŋ/ *noun* [U] all the cables that are required for a particular piece of equipment or a particular system

calculation /ˌkælkjuˈleɪʃn/ *noun* [C,U] finding an answer using mathematics

CAM /kæm/ *abbr* computer-assisted manufacturing; the use of computers to control industrial processes

canal /kəˈnæl/ *noun* [C] an artificial waterway made through land so that boats or ships can travel along it

capacity /kəˈpæsəti/ *noun* [sing, U] the amount that a container or space can hold

cardboard /ˈkɑːdbɔːd/ *noun* [U] very thick paper that is used for making boxes, etc.

casing /ˈkeɪsɪŋ/ *noun* [C, U] a covering that protects sth

Celsius /ˈselsiəs/ *adj* (abbr **C**) of or using a scale of temperature in which water freezes at 0° and boils at 100°

cement /sɪˈment/ *noun* [U] a grey powder that becomes hard after it is mixed with water and left to dry

census /ˈsensəs/ *noun* [C] an official count of the population

centimetre /ˈsentɪmiːtə/ *noun* [C] (abbr **cm**) one hundredth of a metre

centre /ˈsentə/ *noun* [C, usually sing] the middle point or part of sth

certificate /səˈtɪfɪkət/ *noun* [C] an official piece of paper that says that sth is true or correct

chemical engineering /ˌkemikl endʒɪˈnɪərɪŋ/ *noun* [U] the study of the design and use of machines in industrial chemical processes ➤ **chemical engineer** *noun* [C]

chemist /ˈkemɪst/ *noun* [C] a specialist in chemistry

chemistry /ˈkemɪstri/ *noun* [U] the scientific study of the structure of substances and what happens to them in different conditions or when mixed with each other

civil engineering /ˌsɪvl endʒɪˈnɪərɪŋ/ *noun* [U] the design, building and repair of roads, bridges, canals, etc.; the study of this as a subject

classic /ˈklæsɪk/ *adj* important and having a value that will last

clean /kliːn/ *adj* 1 not dirty 2 not showing or having any record of doing sth that is against the law: *a clean driving licence*

clear /klɪə/ *adj* easy to see through (**opposite: opaque**)

clockwork /ˈklɒkwɜːk/ *noun* [U] a type of machinery that you operate by turning a key which winds up a spring and produces energy

CNC /ˌsiː en ˈsiː/ *abbr* computer numerical control; the use of digital computer techniques to control a manufacturing process, especially the machines and tools involved in this process

collapse /kəˈlæps/ *verb* to fall down or break into pieces suddenly

column /ˈkɒləm/ *noun* [C] a tall solid vertical post that supports or decorates a building

combine /ˈkɒmbaɪn/ *verb* to join or mix

communications /kəˌmjuːnɪˈkeɪʃnz/ *noun* [U, also pl] methods of sending information, especially telephones, radio, computers, etc. or roads and railways

component /kəmˈpəʊnənt/ *noun* [C] one of several parts of which sth is made

compound /ˈkɒmpaʊnd/ *noun* [C] something that consists of substances combined together

compress /kəmˈpres/ *verb* to make sth fill less space than usual

compressed-air /kəmˌprest ˈeə/ *adj* using air under pressure as energy to drive machines and tools

computer /kəm'pju:tə/ *noun* [C] an electronic machine that can store, find and arrange information, calculate amounts and control other machines

connect /kə'nekt/ *verb* to join

consistent /kən'sɪstənt/ *adj* always having the same opinions, standard, behaviour, etc.; not changing

construction /kən'strʌkʃn/ *noun* [U] the act or method of making sth or building sth

contact /'kɒntækt/ *noun* [U] (**contact (with sth)**) the state of touching sth

continuous assessment /kən,tɪnjuəs ə'sesmənt/ *noun* [U] a system of giving a student a final mark based on work done during a course of study rather than on one exam

contract /kən'trækt/ *verb* to become or to make sth smaller or shorter

convert /kən'vɜ:t/ *verb* to change sth from one form, system or use to another

cool /ku:l/ *adj* (*slang*) very good or fashionable

copper /'kɒpə/ *noun* [U] a common reddish-brown metal

corner /'kɔ:nə/ *noun* [C] a place where two lines, edges, surfaces or roads meet

corrosive /kə'rəʊsɪv/ *adj* tending to destroy sth slowly by chemical action

course /kɔ:s/ *noun* [C] a complete series of lessons or studies

covering letter /,kʌvərɪŋ 'letə/ *noun* [C] a letter containing extra information

crack /kræk/ *verb* to break or to make sth break so that a line appears on the surface, without breaking into pieces

craftsman /'krɑ:ftsmən/ *noun* [C] a person who makes things skilfully, especially with his/her hands

cubic /'kju:bɪk/ *adj* (abbr **cu**) used to show a measurement of volume (= height × length × width)

cure /kjʊə/ *verb* to make sb healthy again after an illness

curie /'kjʊəri/ *noun* [C] (abbr **Ci**) a unit for measuring radioactivity

curriculum vitae /kə,rɪkjələm 'vi:taɪ/ *noun* [C] (abbr **CV** /,si: 'vi:/) a written record of your education and employment

cycle /'saɪkl/ *noun* [C] the fact of a series of events being repeated many times, always in the same order

D

dam /dæm/ *verb* to build a wall across a river to hold back the water

decimal /'desɪml/ *noun* [C] a number less than one

decimal point /,desɪml 'pɔɪnt/ *noun* [C] a dot or point used to separate a whole number from the tenths, hundredths, etc. of a decimal, for example in **0.61**

defensive /dɪ'fensɪv/ *adj* that protects sb/sth from attack

define /dɪ'faɪn/ *verb* to explain the exact nature or meaning of sth clearly

deflect /dɪ'flekt/ *verb* to change direction after hitting sb/sth; to make sth change direction in this way

degree /dɪ'gri:/ *noun* [C] a measurement of angles

deliver /dɪ'lɪvə/ *verb* to take sth to the place requested

description /dɪ'skrɪpʃn/ *noun* [C] a picture of sb/sth in words

design /dɪ'zaɪn/ *verb* **1** to plan and make a drawing of how sth will be made **2** to invent, plan and develop sth for a particular purpose ➤ **design** *noun* [C]

designer /dɪ'zaɪnə/ *noun* [C] a person whose job is to make drawings or plans showing how sth will be made

desktop /'desktɒp/ *noun* [C] a computer screen on which you can see symbols (**icons**) showing the programs, information, etc. that are available

detail /'di:teɪl/ *noun* [C, U] one fact or piece of information

develop /dɪ'veləp/ *verb* **1** to grow slowly, increase or change into sth else; to make sth do this **2** to think of or produce a new idea, product, etc. and make it successful **3** to begin to have a problem or disease; to start to affect sb/sth ➤ **development** /dɪ'veləpmənt/ *noun* [U,C]

developer /dɪ'veləpə/ *noun* [C] a person or company that designs and creates new products

device /dɪ'vaɪs/ *noun* [C] a tool or piece of equipment

diagonal /daɪ'ægənl/ *adj* (used about a straight line) joining two opposites sides of sth at an angle that is not 90° or vertical or horizontal

diagram /'daɪəgræm/ *noun* [C] a simple picture that is used to explain how sth works or what sth looks like

diameter /daɪ'æmɪtə/ *noun* [C] a straight line that goes from one side to the other of a circle, passing through the centre

diamond /'daɪəmənd/ *noun* [C, U] a hard, bright, precious stone

difficult /'dɪfɪkəlt/ *adj* not easy to do or understand

dimension /dɪ'menʃn/ *noun* [C, U] a measurement in space, for example, the length, width or height of sth ➤ **dimensional** /-ʃnl/ (used to form compound adjectives) having the number of dimensions mentioned

diploma /dɪ'pləʊmə/ *noun* [C] (**a diploma (in sth)**) a certificate for completing a course of study

disadvantage /,dɪsəd'vɑ:ntɪdʒ/ *noun* [C] something that is not good or that causes problems

disaster /dɪ'zɑ:stə/ *noun* [C] an event that causes a lot of harm or damage

disconnect /,dɪskə'nekt/ *verb* **1** to stop a supply of electricity, etc. going to a piece of equipment or a building **2** to separate sth from sth else

disposal /dɪ'spəʊzl/ *noun* [U] the act of getting rid of sth

distance /'dɪstəns/ *noun* [C,U] the space between two places or things

divide /dɪ'vaɪd/ *verb* **divide (sth) (up) (into sth)** to separate (sth) into different parts

drawing board /'drɔ:ɪŋ bɔ:d/ *noun* [C] a board used for holding a piece of paper while a drawing or plan is being made

driving licence /'draɪvɪŋ laɪsns/ *noun* [C] a certificate that proves you have taken a test and are able to drive a car, etc.

duct /dʌkt/ *noun* [C] a tube for carrying liquid, gas, electric or telephone wires

E

ear defenders /'ɪə dɪfendəz/ *noun* [pl] pieces of soft material that you put over your ears to keep out noise

earphones /'ɪəfəʊnz/ *noun* [pl] a piece of equipment that fits over or in the ears and is used for listening to music, the radio, etc.

easy /'i:zi/ *adj* not difficult ➤ **easily** /'i:zəli/ *adv*

efficiently /ɪ'fɪʃntli/ *adv* well and thoroughly with no waste of time, money, or energy

electrical /ɪ'lektrɪkl/ *adj* of or about electricity

electrical engineering /ɪ,lektrɪkl endʒɪ'nɪərɪŋ/ *noun* [U] the design and building of machines and systems that use or produce electricity; the study of this subject

electrical resistance /ɪ,lektrɪkl rɪ'zɪstəns/ *noun* [C,U] the fact of a substance not allowing electricity to flow through it; a measurement of this

electricity /ɪ,lek'trɪsəti/ *noun* [U] a type of energy that we use to make heat, light, and power to work machines, etc.

electrodynamics /ɪ,lektrəʊdaɪ'næmɪks/ *noun* [U] the scientific study of the forces involved in the movement of electricity

electromagnetic /ɪ,lektrəʊmæg'netɪk/ *adj* (in physics) having both electrical characteristics and the ability to attract metal objects

element /'elɪmənt/ *noun* [C] one of the simple chemical substances, eg iron

energy /'enədʒi/ *noun* [U] the power used for driving machines, etc.

engineer /,endʒɪ'nɪə/ *noun* [C] a person whose job is to design, build, or repair engines, machines, etc.

engineering /,endʒɪ'nɪərɪŋ/ *noun* **1** the activity of applying scientific knowledge to the design, building and control of machines, roads, bridges, electrical equipment, etc. **2** [U] the study of this subject

enquiry /ɪn'kwaɪəri/ *noun* [C] **1** a question **2** an official process to find out the cause of sth

environment /ɪn'vaɪrənmənt/ *noun* [C, U] the conditions that affect the behaviour and development of sb/sth; the physical conditions that sb/sth exists in

environmentally-friendly /ɪn,vaɪrən,mentəli 'frendli/ *adj* (used about products) not harming the natural world

equipment /ɪ'kwɪpmənt/ *noun* [U] the things that are needed to do a particular activity

ergonomically /ˌɜːgəˈnɒmɪkli/ *adj* in a way that is designed to help people's working conditions and to help them work more efficiently

estimate /ˈestɪmeɪt/ *verb* to calculate the size, cost, etc. of sth approximately

evaluate /ɪˈvæljueɪt/ *verb* to study the facts and then form an opinion about sth

everyday /ˈevrɪdeɪ/ *adj* used or happening every day or regularly; normal

expand /ɪkˈspænd/ *verb* to become or to make sth bigger

experience /ɪkˈspɪəriəns/ *noun* [U] the things that you have done in your life; the knowledge or skill that you get from seeing or doing sth

expertise /ˌekspɜːˈtiːz/ *noun* [U] a high level of special knowledge or skill

explosive /ɪkˈspləʊsɪv/ *adj* easily able or likely to explode

F

factory /ˈfæktri; ˈfæktəri/ *noun* [C] a building or group of buildings where goods are made by machine;

fantasy /ˈfæntəsi/ *noun* [C,U] a product of your imagination

fashionable /ˈfæʃnəbl/ *adj* popular

fitting /ˈfɪtɪŋ/ *noun* [U] the act of putting or fixing sth somewhere

fixed /fɪkst/ *adj* staying the same; not able to be moved or changed

flammable /ˈflæməbl/ *adj* able or likely to burn easily

flexible /ˈfleksəbl/ *adj* able to bend or move easily without breaking

foil /fɔɪl/ *noun* [U] metal that has been made into very thin sheets, used for putting around food

fold /fəʊld/ *verb* to bend one part of sth over another part in order to make it smaller, tidier, etc. ➤ **fold** *noun* [C]

force /fɔːs/ *noun* 1 [U] physical strength or power 2 [C,U] a power that can cause change or movement

formal /ˈfɔːml/ *adj* serious or official

foundation course /faʊnˈdeɪʃn kɔːs/ *noun* [C] a general course at a college that prepares students for longer or more difficult courses

frequency /ˈfriːkwənsi/ *noun* [C,U] the rate at which a sound wave or radio wave moves up and down (**vibrates**)

fuel consumption /ˈfjuːəl kənsʌmpʃn/ *noun* [U] the act of using heat or power (**fuel**); the amount used

G

gadget /ˈgædʒɪt/ *noun* [C] a small device, tool or machine that has a particular but usually unimportant purpose

generation /ˌdʒenəˈreɪʃn/ *noun* 1 [C] all the people in a family, group or country who were born at about the same time 2 [U] the production of sth, especially heat, power, etc. 3 [C, usually sing] a stage in the development of a product, usually a technical one

generator /ˈdʒenəreɪtə/ *noun* [C] a machine that produces electricity

girder /ˈgɜːdə/ *noun* [C] a long heavy piece of iron or steel that is used in the building of bridges, large buildings, etc.

goggles /ˈgɒglz/ *noun* [pl] special glasses that you wear to protect your eyes from water, wind, dust, etc.

gram /græm/ *noun* [C] (abbr **g**) a measure of weight. There are 1,000 grams in a kilogram.

grind /graɪnd/ *verb* to make sth sharp or smooth by rubbing it on a rough hard surface

guillotine /ˈgɪlətiːn/ *noun* [C] a machine used for cutting paper

H

half /hɑːf/ *noun* [C] (symbol ½) one of two equal parts of sth

hand-drawn /ˈhænd drɔːn/ *adj* drawn or done by a person and not by machine

handle /ˈhændl/ *verb* to touch, hold or move sth with your hands

handmade /ˌhændˈmeɪd/ *adj* made by a person using his/her hands

hard /hɑːd/ *adj* not soft to touch; not easy to break or bend (**opposite: soft**)

hard hat /ˌhɑːd ˈhæt/ *noun* [C] a protective hat

hear /hɪə/ *verb* to receive sounds with your ears ➤ **hearing** /ˈhɪərɪŋ/ *noun* [U]

heavy /ˈhevi/ *adj* weighing a lot; difficult to lift or move (**opposite: light**)

height /haɪt/ *noun* [C,U] the measurement from the bottom to the top of a person or thing

hertz /hɜːts/ *noun* [C] (abbr **Hz**) a unit for measuring the frequency of sound waves

highly-skilled /ˌhaɪli ˈskɪld/ *adj* (used about work, a job, etc.) needing a lot of skills or skill; done by people who have been trained to a high degree

Hoover™ /ˈhuːvə/ *noun* [C] a machine that sucks up the dirt

horizontal /ˌhɒrɪˈzɒntl/ *adj* going from side to side, flat or level

human error /ˌhjuːmən ˈerə/ *noun* [C, U] a mistake made by a person

I

identical /aɪˈdentɪkl/ *adj* exactly the same as; similar in every detail

ignite /ɪgˈnaɪt/ *verb* to start burning or to make sth start burning

illegal /ɪˈliːgl/ *adj* not allowed by the law

image /ˈɪmɪdʒ/ *noun* [C] 1 a copy or picture of sb/sth 2 the general impression that a person, an organization, etc. gives to the public

imagination /ɪˌmædʒɪˈneɪʃn/ *noun* 1 [U,C] the ability to create mental pictures or new ideas 2 [C] the part of the mind that uses this ability

implant /ˈɪmplɑːnt/ *noun* [C] something put into the body in a medical operation

increase /ɪnˈkriːs/ *verb* to become or to make sth larger in number or amount

industrial /ɪnˈdʌstriəl/ *adj* connected with industry

industry /ˈɪndəstri/ *noun* [U] the production of goods in factories

informal /ɪnˈfɔːml/ *adj* relaxed and friendly or suitable for a relaxed occasion

injure /ˈɪndʒə/ *verb* to hurt or harm sb physically

interesting /ˈɪntrəstɪŋ/ *adj* enjoyable and entertaining; holding your attention

interference /ˌɪntəˈfɪərəns/ *noun* [U] extra noise that prevents you from receiving radio, TV, or phone signals clearly

interview /ˈɪntəvjuː/ *noun* [C] a meeting to find out if sb is suitable for a job, course of study, etc.

invention /ɪnˈvenʃn/ *noun* [C] a thing that has been made or designed by sb for the first time ➤ **inventor** /ɪnˈventə/ *noun* [C]

iron /ˈaɪən/ *noun* [U] (symbol **Fe**) a hard strong metal that is used for making steel and is found in small quantities in food and blood

J

Jacuzzi™ /dʒəˈkuːzi/ *noun* [C] a special bath in which powerful movements of air make bubbles in the water

jet /dʒet/ *noun* [C] a fast modern plane

joule /dʒuːl/ *noun* [C] (abbr **J**) a unit of energy or work

K

kelvin /ˈkelvɪn/ *noun* [C,U] (abbr **K**) a unit for measuring temperature

kettle /ˈketl/ *noun* [C] a container with a lid, used for boiling water

key skill /ˌkiː ˈskɪl/ *noun* [C] a particular ability or type of ability

keyboard /ˈkiːbɔːd/ *noun* [C] the set of keys on a computer, etc.

kill /kɪl/ *verb* to make sb/sth die

kilogram /ˈkɪləgræm/ *noun* [C] (abbr **kg**) a measure of weight

kilometre /ˈkɪləmiːtə; kɪˈlɒmɪtə/ *noun* [C] (abbr **k, km**) a measure of length or distance

L

labourer /ˈleɪbərə/ *noun* [C] a person whose job involves hard physical work outdoors

lathe /leɪð/ *noun* [C] a machine that shapes pieces of wood or metal by holding and turning them against a fixed cutting tool

leather /ˈleðə/ *noun* [U] the skin of animals which has been specially treated

length /leŋθ/ *noun* [C] the size of sth from one end to the other

lens /lenz/ *noun* [C] a curved piece of glass that makes things look bigger, clear, etc. when you look through it

light¹ /laɪt/ *adj* not weighing a lot; easy to lift or move (**opposite: heavy**)

light² /laɪt/ *noun* [U,C] the energy from the sun, a lamp, etc. that allows you to see things

lighting /ˈlaɪtɪŋ/ *noun* [U] the quality or type of lights used in a room, etc.

liquid /ˈlɪkwɪd/ *noun* [C,U] a substance, for example water, that is not solid or a gas and that can flow or be poured

listen /'lɪsn/ *verb* to pay attention to sb/sth in order to hear him/her/it

litre /'liːtə/ *noun* [C] (abbr **l**) a measure of liquid

local /'ləʊkl/ *adj* of a particular place (near you)

look /lʊk/ *verb* to turn your eyes in a particular direction (in order to pay attention to sb/sth)

loosen /'luːsn/ *verb* to become or make sth less tight

M

mains /meɪnz/ *noun* [pl] the place where the supply of gas, water or electricity to a building starts; the system of providing these services to a building

maintenance /'meɪntənəns/ *noun* [U] keeping sth in good condition

manufacture /ˌmænjuˈfæktʃə/ *noun* [U] the fact of making sth in large quantities using machines ➤ **manufacturing** /ˌmænjuˈfæktʃərɪŋ/ *noun* [U] ➤ **manufacturer** /ˌmænjuˈfæktʃərə/ *noun* [C]

material /məˈtɪəriəl/ *noun* [C, U] a substance that can be used for making or doing sth

mathematician /ˌmæθəməˈtɪʃn/ *noun* [C] an expert in mathematics

mathematics /ˌmæθəˈmætɪks/ *noun* [U] the science or study of numbers, quantities, or shapes

measurement /'meʒəmənt/ *noun* **1** [C] a size, an amount, etc. that is found by measuring sth **2** [U] the act or process of measuring sth

mechanical engineering /məˈkænɪkl endʒɪˈnɪərɪŋ/ *noun* [U] the study of how machines are designed, built, and repaired

mechanism /'mekənɪzəm/ *noun* [C] a set of moving parts in a machine that does a certain task

medical /'medɪkl/ *adj* connected with medicine and the treatment of illness

memory /'meməri/ *noun* [C, U] **1** a person's ability to remember things **2** the part of a computer where information is stored; the amount of space in a computer for storing information

mend /mend/ *verb* to repair sth that is damaged or broken

metal /'metl/ *noun* [C, U] a type of solid substance that is usually hard and shiny and that heat and electricity can travel through

methodical /məˈθɒdɪkl/ *adj* having or using a well-organized and careful way of doing sth

metre /'miːtə/ *noun* [C] (abbr **m**) a measure of length; 100 centimetres

micrometre /'maɪkrəʊmiːtə/ *noun* [C] one millionth of a metre

mild steel /ˌmaɪld 'stiːl/ *noun* [U] a strong hard material, made from a mixture of iron and carbon

millilitre /'mɪliliːtə/ *noun* [C] (abbr **ml**) one thousandth of a litre

million /'mɪljən/ *noun* [C] 1,000,000

mining /'maɪnɪŋ/ *noun* [U] the process of getting coal and other minerals from under the ground; the industry involved in this

model /'mɒdl/ *noun* [C] a copy of sth that is usually smaller than the real thing

modern /'mɒdn/ *adj* with all the newest methods, equipment, designs, etc.

monitor /'mɒnɪtə/ *noun* [C] a screen that shows information from a computer

mouse /maʊs/ *noun* [C] a piece of equipment, connected to a computer, for moving around the screen and entering commands without touching the keys

movable /'muːvəbl/ *adj* that can be moved (**opposite: fixed**)

N

naturally /'nætʃrəli/ *adv* in a way that is relaxed and normal

newspaper article /'njuːzpeɪpə rɑːtɪkl/ *noun* [C] a piece of writing in a newspaper

newton /'njuːtən/ *noun* [C] (abbr **N**) a unit for measuring force

nickel /'nɪkl/ *noun* [U] (symbol **Ni**) a hard silver-white metal

nuclear /'njuːklɪə/ *adj* using, producing or resulting from the energy that is produced when the central part (**nucleus**) of an atom is split

nylon /'naɪlɒn/ *noun* [U] a very strong man-made material

O

ohm /əʊm/ *noun* [C] (symbol Ω) a unit for measuring electrical resistance

opaque /əʊˈpeɪk/ *adj* that you cannot see through (**opposite: clear**)

opener /'əʊpnə/ *noun* [C] a tool that opens sth or takes the lid, etc. off sth

operator /'ɒpəreɪtə/ *noun* [C] a person whose job is to work a particular machine or piece or equipment

optical fibre /ˌɒptɪkl 'faɪbə/ *noun* [C,U] a thin glass thread through which light can be sent (**transmitted**)

organized /'ɔːgənaɪzd/ *adj* arranged or planned in the way mentioned

original /əˈrɪdʒənl/ *adj* made or created first, before any copies or changes were made

P

parachute /'pærəʃuːt/ *noun* [C] a piece of equipment that is tied to a person and that opens and lets him/her fall to the ground slowly when he/she jumps from a plane

particular /pəˈtɪkjələ/ *adj* used to emphasize that you are talking about one person, thing, time, etc. and not about others

pascal /pæsˈkæl/ *noun* [C] (symbol **Pa**) a unit of pressure equal to one newton per square metre

pass /pɑːs/ *verb* to achieve the necessary standard in an exam, a test, etc.

patience /'peɪʃns/ *noun* [U] the quality of being able to stay calm and not get angry, especially when you have to wait a long time

patient /'peɪʃnt/ *noun* [C] a person who is receiving medical treatment

peaceful /'piːsfl/ *adj* calm and quiet

per /pɜː/ *prep* for each

per cent /pə 'sent/ *noun* [C] (symbol %) one part in every hundred

perfect /'pɜːfɪkt/ *adj* completely good; without faults or weaknesses

personality /ˌpɜːsəˈnæləti/ *noun* [C,U] the different qualities of a person's character that make him/her different from other people

petrochemical /ˌpetrəʊˈkemɪkl/ *noun* [C] any chemical substance obtained from petroleum oil or natural gas

petroleum /pəˈtrəʊliəm/ *noun* [U] mineral oil that is found under the ground or the sea and is used to produce petrol

physicist /'fɪzɪsɪst/ *noun* [C] an expert in physics

physics /'fɪzɪks/ *noun* [U] the scientific study of natural forces

pie chart /'paɪ tʃɑːt/ *noun* [C] a diagram consisting of a circle divided into parts to show the size of particular parts in relation to the whole

pier /'pɪə/ *noun* [C] a large structure that is built in the sea or a river to support a bridge where it crosses the water

pilot /'paɪlət/ *noun* [C] a person who flies an aircraft

pioneer /ˌpaɪəˈnɪə/ *noun* [C] a person who is one of the first to develop an area of human knowledge, culture, etc.

plant /plɑːnt/ *noun* [C] a factory or place where power is produced or where an industrial process takes place

plastic /'plæstɪk/ *noun* [C, U] a light, strong material that is made with chemicals and is used for making many different sorts of objects ➤ **plastic** *adj*

plug /plʌg/ *noun* [C] sth which connects a piece of electrical equipment to the electricity supply

plug sth in *phrasal verb* to connect a piece of electrical equipment to the electricity supply or to another piece of equipment

point /pɔɪnt/ *noun* [C] the thin sharp end of sth

polymer /'pɒlɪmə/ *noun* [C] a compound consisting of large groups of atoms (**molecules**) made from combinations of small simple molecules

population /ˌpɒpjuˈleɪʃn/ *noun* [C,U] the number of people who live in a particular area, city or country

power[1] /'paʊə/ *noun* [U] energy that can be collected and used for operating machines, making electricity, etc.

power[2] /'paʊə/ *verb* to supply energy to sth to make it work

practical /'præktɪkl/ *adj* **1** concerned with actually doing sth rather than with ideas or thought **2** very suitable for a particular purpose; useful

precision /prɪˈsɪʒn/ *noun* [U] the quality of being clear or exact

prepare /prɪˈpeə/ *verb* to get ready or to make sb/sth ready

prevent /prɪˈvent/ *verb* (**prevent sb/sth (from) (doing sth)**) to stop sth happening or to stop sb from doing sth

printer /ˈprɪntə/ *noun* [C] a machine that prints out information from a computer onto paper

process /ˈprəʊses/ *noun* [C] a series of actions that you do for a particular purpose

produce /prəˈdjuːs/ *verb* **1** to make sth to be sold, especially in large quantities **2** to cause a particular effect or result ► **production** /prəˈdʌkʃn/ *noun* [C]

project /ˈprɒdʒekt/ *noun* [C] a piece of work that is planned and organized carefully

property /ˈprɒpəti/ *noun* [C, usually pl] a special quality or characteristic

proportion /prəˈpɔːʃn/ *noun* **1** [C] a part or share of a whole **2** [U] the relationship between the size or amount of two things

prototype /ˈprəʊtətaɪp/ *noun* [C] the first model or design of sth from which other forms will be developed

put sth together *phrasal verb* to build or repair sth by joining its parts together

Q

qualification /ˌkwɒlɪfɪˈkeɪʃn/ *noun* [C, usually pl] an exam that you have passed or a course of study that you have successfully completed

quarter /ˈkwɔːtə/ *noun* [C] (symbol ¼) one of four equal parts of sth

R

radioactive /ˌreɪdiəʊˈæktɪv/ *adj* sending out powerful and very dangerous rays that are produced when atoms are broken up. These rays cannot be seen or felt but can cause serious illness or death. ► **radioactivity** /ˌreɪdiəʊækˈtɪvəti/ *noun* [U]

react /riˈækt/ *verb* (used about a chemical substance) to change after coming into contact with another substance

realistically /ˌrɪəˈlɪstɪkli/ *adv* in a sensible and understanding way

receiver /rɪˈsiːvə/ *noun* [C] a person who gets a letter, a message, etc. from sb

recycle /ˌriːˈsaɪkl/ *verb* **1** to put used objects and materials through a process so that they can be used again **2** to keep used objects and materials and use them again

referee /ˌrefəˈriː/ *noun* [C] a person who gives information about your character and ability, usually in a letter, for example when you are hoping to be chosen for a job

register /ˈredʒɪstə/ *verb* to show sth or be shown on a measuring instrument

relaxed /rɪˈlækst/ *adj* not worried or tense

reliable /rɪˈlaɪəbl/ *adj* that you can trust

repair /rɪˈpeə/ *verb* to put sth old or damaged back into good condition

repeater /rɪˈpiːtə/ *noun* [C] a device which automatically transmits or sends again an electronically transmitted message

repetitive strain injury /rɪˌpetətɪv ˈstreɪn ɪndʒəri/ *noun* [U] (abbr **RSI** /ˌɑːr es ˈaɪ/) pain and swelling, especially in the wrists and hands, caused by doing the same movement many times

require /rɪˈkwaɪə/ *verb* to need sth

resource /rɪˈzɔːs/ *noun* [C, usually pl] a supply of sth, a piece of equipment, etc. that is available for sb to use

rigid /ˈrɪdʒɪd/ *adj* difficult to bend or shape; stiff

risk /rɪsk/ *noun* [C,U] a possibility of sth dangerous or unpleasant happening; a situation that could be dangerous or have a bad result

robot /ˈrəʊbɒt/ *noun* [C] a machine that works automatically or is controlled by a computer

robotics /rəʊˈbɒtɪks/ *noun* [U] the science of designing and operating robots

rust /rʌst/ *verb* to become covered with a reddish-brown substance which forms on the surface of iron, etc. and is caused by the action of air and water

S

safe /seɪf/ *adj* not likely to cause danger, harm or risk ► **safely** /ˈseɪfli/ *adv* ► **safety** /ˈseɪfti/ *noun* [U]

sanitation /ˌsænɪˈteɪʃn/ *noun* [U] the equipment and the systems that keep places clean, especially by removing human waste

scale /skeɪl/ *noun* [C] a series of numbers amounts, etc. that are used for measuring or fixing the level of sth

scientific /ˌsaɪənˈtɪfɪk/ *adj* connected with or involving science ► **scientist** /ˈsaɪəntɪst/ *noun* [C]

screw /skruː/ *noun* [C] a thin pointed piece of metal used for fixing two things together. You turn a screw with a special tool (**a screwdriver**).

seabed /ˈsiːbed/ (**the seabed**) *noun* [sing] the floor of the sea

secret /ˈsiːkrət/ *adj* known about by only a few people; kept hidden from others

secretary /ˈsekrətri/ *noun* [C] a person who works in an office. A secretary types letters, answers the telephone, keeps records, etc.

see /siː/ *verb* to become conscious of sth using your eyes

sender /ˈsendə/ *noun* [C] a person who sends a letter, a package, etc. to sb

sense¹ /sens/ *noun* [C] one of the five natural physical powers of sight, hearing, smell, taste and touch

sense² /sens/ *verb* to realize or become conscious of sth; to get a feeling about sth you cannot see, hear, etc.

sensible /ˈsensəbl/ *adj* (used about people and their behaviour) able to make good judgements based on reason and experience; practical

sensor /ˈsensə/ *noun* [C] a device that can react to light, heat, pressure, etc.

setting /ˈsetɪŋ/ *noun* [C] one of the positions of the controls of a machine

shape¹ /ʃeɪp/ *verb* to make sth into a particular form

shape² /ʃeɪp/ *noun* [C, U] the form of the outer edges or surfaces of sth; an example of sth that has a particular form

shape memory /ˈʃeɪp meməri/ *adj* (used about a substance or material) able to change and adapt according to its surroundings

side /saɪd/ *noun* [C] one of the surfaces of sth except the top, bottom, front or back

sight /saɪt/ *noun* [U] the ability to see

signal /ˈsɪgnəl/ *noun* [C] a series of radio waves, etc. that are sent out or received

skill /skɪl/ *noun* **1** [U] the ability to do sth well, especially because of training, practice, etc. **2** [C] an ability that you need in order to do a job, an activity, etc. well ► **skilled** *adj*

sloping /ˈsləʊpɪŋ/ *adj* (used about a surface) not flat: built at an angle

smart /smɑːt/ *adj* **1** (used about a piece of clothing) formal **2** clever; intelligent **3** fashionable

smell /smel/ *noun* [U] the ability to sense things with the nose

socket /ˈsɒkɪt/ *noun* [C] **1** a place in a wall where a piece of electrical equipment can be connected to an electrical supply **2** a hole in a piece of electrical equipment where another piece of electrical equipment can be connected

soft /sɒft/ *adj* not hard or firm

software /ˈsɒftweə/ *noun* [U] the programs and other operating information used by a computer

solar energy /ˌsəʊlə ˈenədʒi/ *noun* [U] power from the sun

solution /səˈluːʃn/ *noun* [C] a way of finding the answer to a problem or dealing with a difficult situation

solve /sɒlv/ *verb* to find a way of dealing with a problem or difficult situation

span /spæn/ *noun* [C] the length of sth from one end to the other

specialist /ˈspeʃəlɪst/ *noun* [C] an expert in a particular area

specialize /ˈspeʃəlaɪz/ *verb* (**specialize (in sth)**) **1** to become an expert in a particular area of work or study **2** to give most of your attention to one subject, type of product, etc.

speed /spiːd/ *noun* [C,U] the rate at which sb/sth moves or travels

splicing /ˈsplaɪsɪŋ/ *noun* [U] the act of joining the ends of two pieces of cable, etc. together

square /skweə/ *adj* (abbr **sq**) used after a number to give a measurement of an area

stage /steɪdʒ/ *noun* [C] one part of the progress or development of sth

state-of-the-art /ˌsteɪt əv ði ˈɑːt/ *adj* using the most modern or advanced techniques and methods

statistics /stəˈtɪstɪks/ *noun* [pl] numbers that have been collected in order to provide information about sth

Stetson™ /'stetsən/ *noun* [C] a type of hat typical in Texas, USA

stitch /stɪtʃ/ *noun* [C] one of the small pieces of thread that a doctor uses to sew your skin together after an operation, etc.

strain /streɪn/ *noun* 1 [C,U] pressure or worry 2 [U] pressure that is put on sth when it is pulled or pushed by a physical force

strength /strenθ/ *noun* [U] the ability of an object to hold heavy weights or not to break or be damaged easily
➤ **strengthen** /'strenθn/ *verb* to make sth stronger

strong /strɒŋ/ *adj* (used about a thing) not easily broken or damaged (**opposite: weak**)

structural /'strʌktʃərəl/ *adj* connected with the way that a building, etc. has been built or the way that the parts of sth have been put together

substandard /səb'stændəd/ *adj* not as good as normal; not acceptable

suction /'sʌkʃn/ *noun* [U] the action of removing air from a space or container so that two surfaces can stick together

supply¹ /sə'plaɪ/ *noun* [C] a store or an amount of sth that is provided or available to be used

supply² /sə'plaɪ/ *verb* to give or provide sth

surgery /'sɜːdʒəri/ *noun* [U] medical treatment in which your body is cut open so that part of it can be removed or repaired

survey /'sɜːveɪ/ *noun* [C] a study of the opinions, behaviour, etc. of a group of people

suspension bridge /sə'spenʃn brɪdʒ/ *noun* [C] a bridge that hangs from thick steel wires that are supported by towers at each end

symbol /'sɪmbl/ *noun* [C] a sign, object, etc. that represents sth

symmetrical /sɪ'metrɪkl/ *adj* having two halves that match each other exactly in size, shape, etc.

symptom /'sɪmptəm/ *noun* [C] a change in your body that is a sign of illness

system /'sɪstəm/ *noun* [C] a set of ideas or rules for organizing sth; a particular way of doing sth

T

taste¹ /teɪst/ *noun* [U] the ability to recognize the flavour of a food or drink
➤ **taste²** /teɪst/ *verb*

technical /'teknɪkl/ *adj* connected with the practical use of machines, methods, etc. in science or industry

technician /tek'nɪʃn/ *noun* [C] a person whose job is keeping a particular type of equipment or machinery in good condition

technique /tek'niːk/ *noun* [C] a particular way of doing sth

technology /tek'nɒlədʒi/ *noun* 1 [U,C] scientific knowledge, used in practical ways in industry 2 [C,U] the scientific knowledge and/or equipment that is needed for a particular industry, etc.

teeth brace /'tiːθ breɪs/ *noun* [C, often pl] a metal frame that is fixed to a child's teeth in order to make them straight

telecommunications /ˌtelɪkəˌmjuːnɪ'keɪʃns/ *noun* [pl] the technology of sending signals, images and messages over long distances by radio, telephone, television, etc.

temperature /'temprətʃə/ *noun* [C, U] how hot or cold sb/sth is

tension /'tenʃn/ *noun* [C,U] the condition of not being able to relax because you are worried or nervous

test /test/ *verb* to use a machine, product, etc. to find out how well it works
➤ **tester** /'testə/ *noun* [C]

textile /'tekstaɪl/ *noun* [C, U] any cloth made in a factory

theoretical /ˌθɪə'retɪkl/ *adj* based on ideas and principles, not on practical experience

thermometer /θə'mɒmɪtə/ *noun* [C] an instrument for measuring temperature

think /θɪŋk/ *verb* to use your mind to consider sth or to form connected ideas

third /θɜːd/ *noun* [C] (symbol $^1/_3$) one of three equal parts of sth

thousand /'θaʊznd/ *noun* [C] 1,000

thread /θred/ *noun* [C] a long thin piece of cotton, wool, etc. that you use for sewing, etc.

tighten /'taɪtn/ *verb* to make sth tight or tighter

titanium /tɪ'teɪniəm/ *noun* [U] (symbol Ti) a silver-white metal that is used in making various strong light materials

tomb /tuːm/ *noun* [C] a large place, where the body of an important person is buried

ton /tʌn/ *noun* [C] a measure of weight; 2,240 pounds

tool /tuːl/ *noun* [C] an instrument that you hold in your hand and use for making or repairing things, etc.

top /tɒp/ *noun* [C] the highest part or point of sth

touch /tʌtʃ/ *noun* [U] the ability to feel things and know what they are like by putting your hands or fingers on them
➤ **touch** /tʌtʃ/ *verb*

tough /tʌf/ *adj* not easy to break or damage

tour /tɔː/ *noun* [C] a short visit around a building, city, etc.

traditional /trə'dɪʃənl/ *adj* following the beliefs, customs or way of life of a group of people that have not changed for a long time

transmission /tæns'mɪʃn/ *noun* [U] sending out or passing sth from one person, place or thing to another

U

under /'ʌndə/ *prep, adv* in or to a position that is below sth

unique /ju'niːk/ *adj* not like anything else; being the only one of its type

unplug /ˌʌn'plʌg/ *verb* to remove a piece of electrical equipment from the electricity supply

unreliable /ˌʌnrɪ'laɪəbl/ *adj* that cannot be trusted or depended on to work properly

upholsterer /ʌp'həʊlstərə/ *noun* [C] a person whose job is to cover furniture with soft material and fabric

useful /'juːsfl/ *adj* having some practical use; helpful

V

valve /vælv/ *noun* [C] a device in a pipe or a tube which controls the flow of air, liquid or gas, letting it move in one direction only

vehicle /'viːəkl/ *noun* [C] something that transports people or things

vertical /'vɜːtɪkl/ *adj* going straight up at an angle of 90° from the ground

viewer /'vjuːə/ *noun* [C] a person who looks at or considers sth

vinyl /'vaɪnl/ *noun* [C,U] a strong plastic that can bend easily and is used for making wall, floor and furniture coverings

vision /'vɪʒn/ *noun* [U] the ability to see; sight

volt /vəʊlt/ *noun* [C] (abbr V) a unit for measuring the force of an electric current

W

waiting list /'weɪtɪŋ lɪst/ *noun* [C] a list of people who are waiting for sth, for example a service or medical treatment, that will be available in the future

waterway /'wɔːtəweɪ/ *noun* [C] a canal, river, etc. along which boats can travel

watt /wɒt/ *noun* [C] (abbr W) a unit for measuring electrical power

weak /wiːk/ *adj* that cannot support a lot of weight; likely to break (**opposite: strong**)

weight /weɪt/ *noun* [C,U] how heavy sb/sth is; the fact of being heavy

welder /'weldə/ *noun* [C] a person whose job is joining pieces of metal together by heating their edges and pressing them together ➤ **welding** /'weldɪŋ/ *noun* [U]

welding torch /'weldɪŋ tɔːtʃ/ *noun* [C] a tool with a very hot flame that is used to join pieces of metal together

wind /waɪnd/ *verb* (**wind sth up**) to make a clock or other mechanism work by turning a key, a handle, etc. several times ➤ **wind-up** *adj* that can be make to work by being wound up

wire /'waɪə/ *noun* [C] a piece of metal in the form of a thin thread that is used to carry electricity

wiring /'waɪərɪŋ/ *noun* [U] the system of wires which supplies electricity to rooms in a building

working environment /'wɜːkɪŋ ɪn'vaɪrənmənt/ *noun* [C,U] the conditions that sb works in

Z

zero /'zɪərəʊ/ *noun* [C] the number 0

OXFORD
UNIVERSITY·PRESS

Great Clarendon Street, Oxford OX2 6DP

Oxford University Press is a department of the University of Oxford. It furthers the University's objective of excellence in research, scholarship, and education by publishing worldwide in

Oxford New York

Auckland Bangkok Buenos Aires Cape Town Chennai Dar es Salaam Delhi Hong Kong Istanbul Karachi Kolkata Kuala Lumpur Madrid Melbourne Mexico City Mumbai Nairobi São Paulo Shanghai Taipei Tokyo Toronto

Oxford and Oxford English are registered trade marks of Oxford University Press in the UK and in certain other countries

ISBN 0 19 438827 1

Printed in Spain by Unigraf S.L.

Acknowledgements

The authors and publishers are very grateful to Norman Glendinning for specialist Engineering advice and to the teachers who read and piloted material and gave invaluable feedback.

Geert Claeys, Pavla Čípová, Höfle Lászlóné, Hana Sedláková, Alena Šteklova

We would like to thank the following for their kind permission to reproduce photographs and other copyright material:
Trevor Baylis p 18
BMW p10 (2002 Mini)
Peter Grant p28
Innovations p20
Morgan p11
Phillips p13
Photodisc royalty free pp2, 4, 6, 8, 9, 14, 19, 24, 26, 27, 30
QA Photos Ltd p17 (J Byrne)
Rex Features p10 (1960s Mini)
Mike Rowland/Clifton Suspension Bridge Visitor Centre p22
Science Photo Library p16

Illustrations by:
Hardlines pp12, 13, 15, 22, 23
Keith Shaw pp25, 29

Commissioned photography by:
Phil James p12